# AI Shaping the Future

## How Artificial Intelligence Is Transforming Innovation, Humanity and Ethics

Javad Yahaghi

Javad Yahaghi

## Acknowledgments

Writing this book has been a transformative journey, and it would not have been possible without the support, encouragement, and insights of many people. I want to take this moment to extend my sincere thanks to those who played a vital role.

First, to my wife, Anne—thank you for your unwavering belief in me, your patience, and your love throughout the long writing process. To my children, Atrisa and Artin—your curiosity and joy are constant reminders of why shaping the future is so important.

To my parents, who always believed in my visions and encouraged me to pursue them with passion—thank you for your unconditional support.

To my YouTube followers from @AIFutureHub and the @NovAugment community, your enthusiasm and insightful questions inspired many ideas in this book.

Special thanks to the researchers, educators, and visionaries in the field of AI whose work guided this exploration. I also acknowledge my mentors and peers who offered valuable suggestions, keeping this book accurate and engaging.

Finally, this book is dedicated to everyone who believes in the potential of technology not only to innovate but to uplift humanity. Thank you all for being a part of this project.

# Table of Contents:

# **Introduction**

## Why This Book?

We live in a time when artificial intelligence (AI) technologies influence nearly every dimension of our lives often in ways we may not even notice. From personalized recommendations on streaming services to AI-powered language translation, from autonomous vehicles to sophisticated medical diagnostics, AI systems have quickly become integral to both our daily routines and the global economy. Yet many people's impressions of AI are shaped by media speculation, sensationalist headlines, or dystopian fiction. As a result, misconceptions abound, and legitimate questions about the ethical, social, and economic impact of AI remain under-explored in mainstream conversations.

This book endeavors to offer clarity and depth, providing a thorough exploration of how AI works, how it transforms various industries, and how it intersects with larger societal issues like governance, ethics, and even the possibility of machine consciousness. It also aims to offer pragmatic guidance on how citizens, professionals, and policy makers can shape AI's future rather than passively watch from the sidelines.

## Reshaping Our World

In recent years, we've entered what many describe as the "Fourth Industrial Revolution," in which the lines between our physical, digital, and biological worlds begin to blur. AI stands squarely in the center of this revolution. Whether it's a machine learning model that predicts stock prices for institutional investors or an intelligent tutor that adapts to each student's learning style, AI's potential to remake our core economic and social infrastructures is vast.

Simultaneously, we face a convergence of global challenges climate change, pandemics, socio-economic inequality, data privacy, and more that demand new forms of innovation. AI, if developed thoughtfully, could be a linchpin in tackling these grand challenges. But if it's deployed without adequate safeguards, it could exacerbate inequalities, erode civil liberties, and concentrate power in the hands of a few. The stakes couldn't be higher.

## My Personal Journey and Vision

My own journey with AI began when I was a PhD student fascinated by the possibility of teaching machines to recognize speech and images. Over time, I moved on to deep learning, reinforcement learning, and, ultimately, the question of how machines might one day approach human-level intelligence or even surpass it. This trajectory taught me that AI progress is neither strictly linear nor predictable: breakthroughs sometimes spring from unexpected corners of research, and technologies can leap from academic prototypes to commercial juggernauts in a matter of months.

Amid this excitement, I've also seen the sobering limitations of AI systems. Failures in algorithmic fairness, risks to autonomy and privacy, and unforeseen security vulnerabilities are all very real. When it comes to forging the future of AI, neither utopian nor dystopian extremes provide the whole picture. Instead, I believe we must engage critically with the full complexity technical, social, economic, ethical to ensure AI serves the greater good.

## A Roadmap of What's Ahead

In this book, you'll find five parts, each deeply explored:

### *Foundations of AI*

Chapters 1 - 3 delve into what AI is, how it evolved, how machine learning differs from traditional software, and the elusive dream of Artificial General Intelligence (AGI).

### *Transforming Industries, Transforming Lives*

Chapters 4 - 7 survey AI's impact on healthcare, education, finance, transportation, agriculture, and the creative arts illustrating both concrete benefits and emerging challenges.

### *Ethics, Governance, and Consciousness*

Chapters 8 - 11 explore ethical frameworks, policy debates, existential risks, and the theoretical threshold of AI that might "think" or "feel."

### *The Next Frontier of Innovation*

Chapters 12 - 14 spotlight how AI could reshape the future via smart cities, sustainability, biotechnology, quantum computing, aerospace, and more.

### *Empowering the Individual from Consumer to Creator*

Chapters 15 - 18 focus on you, the reader: learning pathways, civic engagement, new professions, and strategies to ensure technology remains a tool for human flourishing.

Throughout these sections, you will find historical context, technical explanations, vivid use cases, ethical considerations, and forward-looking insights. Let us embark on this journey together, maintaining open minds and a readiness to confront both the daunting and inspiring possibilities of AI.

Javad Yahaghi

# Part 1: Understanding the Foundations of AI

# Chapter 1: Demystifying AI

## A Brief Historical Arc of AI

Modern AI research dates back to the 1950s, when pioneers like Alan Turing and John McCarthy pondered whether machines could simulate human intelligence. In 1956, the Dartmouth Conference formally coined the term "Artificial Intelligence," setting the stage for decades of exploration. Early successes included programs capable of proving mathematical theorems or playing checkers at a high level. Yet progress stalled during the "AI winters" of the 1970s and late 1980s, when inflated promises failed to meet technological realities, and research funding dried up.

Although these winters often led to a lull in public excitement and investment, they also acted as filters only the most resilient research efforts continued to push the boundaries of what AI could accomplish. Over time, these dedicated researchers refined theoretical foundations, improved computational methods, and formed interdisciplinary teams that fostered synergy between computer science, mathematics, and cognitive psychology. This cross-pollination of ideas laid the groundwork for several modern breakthroughs, reminding us that AI is not just about algorithms but about the dynamic interaction between human ingenuity, data availability, and evolving hardware capabilities.

A pivotal shift came in the mid-2000s with the confluence of three factors:

**Data Explosion:** The internet, social media, and digitized records generated massive datasets for training.

**Computational Power:** GPUs (initially for gaming) and specialized hardware enabled parallel processing of vast amounts of data.

**Algorithmic Innovation:** Neural networks especially deeper, multi-layered networks returned to the spotlight with better training techniques, catalyzing "deep learning."

During this period, researchers also realized that collecting higher-quality data was just as crucial as gathering large volumes of it. Academics and industry practitioners alike began curating specialized datasets such as ImageNet for computer vision enabling more precise and reliable experiments. Parallel to this, the global tech community benefited from open-source frameworks like TensorFlow and PyTorch, which dramatically lowered the barrier to entry for novices and advanced developers. These new, accessible tools accelerated innovation, spurring collaborations that transcended academic labs and gave rise to AI startups targeting domains as varied as healthcare diagnostics, natural language processing, and robotics.

Moreover, the resurgence of neural networks coincided with breakthroughs in optimization techniques like improved gradient-based methods that made it feasible to train increasingly deep architectures. Instead of focusing on narrow, rule-based systems, the AI field shifted to exploring large-scale approaches that leverage massive computational clusters and cloud services. This set the stage for cutting-edge research into areas like reinforcement learning, generative modeling, and self-supervised learning, all of which further expanded the horizons of AI's capabilities.

## Core Terminologies in AI

AI: A broad field aiming to create machines that can perform tasks requiring human-like intelligence reasoning, perception, learning, and problem-solving.

Machine Learning (ML): A subset of AI focusing on algorithms that improve through experience (data). This category includes supervised learning, unsupervised learning, and reinforcement learning.

Deep Learning: A subfield of ML that uses multilayered neural networks to automatically learn abstract representations from data (e.g., identifying features like edges in images or semantics in text).

Neural Networks: Models loosely inspired by biological neurons. They consist of layers of "artificial neurons" that pass numerical signals forward (and sometimes backward, via backpropagation) to learn patterns.

It is important to distinguish between these core concepts because each plays a unique role in shaping AI research and its real-world applications. Traditional AI (often associated with "symbolic AI") focuses on predefined rules and logic, whereas machine learning (ML) thrives on large datasets, automatically extracting patterns. Deep learning specifically excels at handling complex tasks like image recognition by layering multiple representations of the data, each building upon the previous layer's output. Understanding these distinctions not only clarifies how AI systems work but also highlights what each approach can and cannot do.

For instance, while traditional ML requires carefully engineered "features" (like color histograms for images), deep learning automates the feature extraction process, often achieving remarkable performance gains. Nevertheless, these powerful techniques come with drawbacks: they can be computationally expensive, data-hungry, and sometimes difficult to interpret. As a result, selecting the right approach symbolic AI, ML, or deep learning depends heavily on the specific problem, data availability, and resource constraints.

## Why AI Surged: The Fourth Industrial Revolution

Some refer to the age of AI as the Fourth Industrial Revolution preceded by steam power, electricity, and digital computing. Today's revolution merges digital (computers, data analysis), physical (robotics, autonomous systems), and biological (genomics, neurotechnology) domains into a seamless ecosystem. AI's ability to spot patterns at scale supercharges these converging fields, forging new capabilities, industries, and ethical dilemmas, such as data privacy and algorithmic fairness.

From customizing e-commerce recommendations to identifying galaxies in vast astronomical data sets, AI's core logic learning from examples unlocks unprecedented possibilities. Yet, the same power also sparks concerns over surveillance, monopolistic data control, and the displacement of labor. Therefore, understanding AI's basis and trajectory is critical for guiding it responsibly.

In addition to reshaping traditional industries like manufacturing and finance, AI is fueling the emergence of entirely new markets such as data annotation services, personalized digital health platforms, and even AI-driven art marketplaces. Policy-makers worldwide are scrambling to craft regulations that balance innovation with ethical considerations. We see debates about algorithmic transparency, data privacy, and workforce retraining in parliaments, boardrooms, and living rooms alike.

It's also worth noting that AI amplifies our ability to solve global challenges. Machine learning models are increasingly used to forecast natural disasters, optimize renewable energy grids, and accelerate vaccine development. Scientists and engineers collaborate on AI-driven climate models that offer unprecedented accuracy in predicting extreme weather events. In medicine, AI helps analyze vast gene-editing data to uncover new treatments for hereditary diseases. These breakthroughs, while awe-inspiring, also highlight our responsibility to ensure that AI's benefits are broadly distributed and do not exacerbate existing inequalities.

On the flip side, the data-centric nature of AI means that organizations with the largest datasets and strongest computational infrastructure often tech giants wield disproportionate power. This dynamic fuels discussions about antitrust regulations, data monopolies, and the ethical implications of data ownership. As we move further into this Fourth Industrial Revolution, the central question becomes how to harness AI's innovations without sacrificing democratic values, personal liberties, or economic fairness.

# Chapter 2: From Code to Cogitation: How AI Differs from Traditional Software

## The Traditional Software Paradigm

Before the rise of machine learning, software was almost entirely "rule-based." Engineers would specify the exact logic: if X happens, do Y. This made sense for well-defined tasks (e.g., arithmetic operations, basic databases) but quickly hit limitations in more complex domains (e.g., image recognition, natural language understanding), where enumerating every rule was either impossible or prohibitively expensive.

Early rule-based systems gained traction in fields like accounting or inventory management, where the data followed strict formats and the workflows remained largely predictable. However, the moment software developers attempted to handle nuanced tasks such as processing natural language sentences or spotting anomalies in images the "if-then" structures became unwieldy. Each additional rule introduced new edge cases, and the complexity could balloon so fast that maintenance became a logistical nightmare.

Moreover, traditional software paradigms assumed that human experts could accurately capture all the relevant knowledge in a given domain. This meant spending countless hours interviewing specialists, codifying their expertise into rule sets, and then testing these rules against real-world scenarios. If the domain was well-understood like basic math or accounting this strategy worked efficiently. But for dynamic, ambiguous, or highly variable domains (think facial recognition, medical diagnoses, or nuanced language interpretation), even the most detailed rules struggled to keep pace with real-world variability.

This gap between "human-defined rules" and "infinite real-world variety" was the Achilles' heel of traditional software approaches. As data volumes grew in the 1990s and early 2000s, it became painfully clear that we needed a new paradigm one that could adapt and learn from the data itself, rather than relying solely on painstakingly programmed logic. Researchers began to explore ways to make computers infer patterns on their own, laying the groundwork for machine learning and, eventually, deep learning.

## The Learning Paradigm

Machine learning flips the traditional approach. Instead of explicitly coding the solution, developers provide a large labeled or unlabeled dataset, and the system learns patterns automatically:

Data Collection: The data could be images, text, or sensor readings from Internet of things (IoT) devices.

Training: The ML algorithm iterates over the data, adjusting its internal parameters (weights, biases) to reduce error. Techniques like gradient descent fine-tune these parameters.

Generalization: The model applies what it learned to new, unseen data recognizing objects or predicting outcomes.

This approach is powerful but also more opaque. Understanding precisely "how" the machine arrives at its decisions can be challenging (hence the notion of the "black box" problem). Additionally, if the training data is biased or flawed, the AI's outputs will reflect and amplify those flaws raising ethical and legal questions about responsibility and accountability.

In essence, machine learning shifts the burden of detail from the programmer to the algorithm, which scours vast quantities of data to "learn" the underlying relationships. For a simple example, consider a spam filter. Instead of a developer painstakingly crafting thousands of rules for spotting spammy keywords or suspicious email structures, a machine learning model simply ingests hundreds of thousands of emails labeled as "spam" or "not spam." It then discerns which word frequencies, sender patterns, or message lengths are most predictive of spam. This automated discovery of patterns often outperforms manual rule sets, especially as spam tactics evolve.

Crucially, the success of machine learning hinges on the quality and diversity of the training data. If the dataset does not represent the full spectrum of real-world examples for instance, it omits emails in certain languages or from certain regions the model may fail to recognize legitimate messages or properly flag spam for those user groups. This highlights a fundamental principle in AI: "garbage in, garbage out." If you feed a model with biased or unrepresentative data, the outcomes will mirror or even amplify those biases.

Despite these challenges, the learning paradigm powers most of today's impressive AI achievements. Speech recognition systems like those found in virtual assistants (Siri, Alexa, Google Assistant) rely on massive audio datasets to capture the nuances of human speech across diverse accents. Image classifiers, used in everything from social media tagging to medical diagnostics, are trained on millions of labeled pictures. Even generative AI models that craft textual or visual content draw from massive corpora spanning the internet, learning to emulate different writing styles, artistic aesthetics, and domain-specific vocabularies.

However, this remarkable adaptability comes at a cost: explainability. Neural networks, in particular, employ layers of computation that transform raw inputs (like pixel values) into higher-level representations (like identifying the shape of an eye in a face) before making a final prediction. Each layer's workings can be hard to interpret. While researchers are developing methods for visualizing what neural networks "see" and which input features they rely on, there's still a broad consensus that these models can be opaque. This opacity spurs debates in industries that require rigorous audit trails like finance, healthcare, or autonomous vehicles where "the machine said so" is insufficient to justify critical decisions.

## Continuous Improvement and Limitations

AI systems can continue learning post-deployment (online learning). For example, a spam filter refines its accuracy as more examples of spam emails appear. But this also introduces "dataset drift" if new malicious actors emerge using methods not seen in the training data, the model's performance can degrade until it is retrained.

Limitations include the heavy reliance on large, high-quality datasets, computational costs of model training, and challenges in interpretability. These constraints highlight that despite the excitement around AI, its real-world success depends on thoughtful data curation, robust model architectures, and human oversight.

The idea of AI systems learning continuously in production has transformed entire industries. Netflix and YouTube recommender systems monitor user behavior what videos are watched, liked, or skipped to fine-tune algorithms on the fly. Similarly, ride-sharing apps monitor traffic patterns and surge pricing data in real time to optimize routes and fares. Yet, this constant adaptation can become a double-edged sword. If a malicious actor manipulates input data by orchestrating fake reviews, clicks, or ratings a continuously learning system could embed these distortions, yielding skewed outputs.

Moreover, dataset drift can occur not just from malicious activity but also from natural evolutions in user behavior or external events. Consider a voice-recognition model that was trained primarily on speech patterns from urban environments. Over time, as more rural users adopt the system, the model may struggle to understand different dialects or background noises, undermining performance. Regular retraining and targeted data collection strategies become essential for maintaining accuracy.

From a computational standpoint, training cutting-edge AI models can be incredibly resource-intensive. Large neural networks with billions of parameters often require powerful GPUs or specialized hardware (like TPUs) and can take days or weeks to train, even in well-funded research labs. This leads to concerns about energy consumption and the environmental footprint of AI. Some in the field champion more efficient model architectures or "green AI" initiatives, which aim to reduce the computational overhead while preserving performance levels.

Interpretability remains another major bottleneck. In high-stakes applications such as medical diagnoses or loan approvals understanding how an AI system arrived at a decision is critical for accountability and trust. Policymakers, ethicists, and researchers are collaborating to develop "explainable AI" techniques, methods that offer insights into which factors influenced a model's output. For instance, "local explanation" tools might highlight words in a text or regions in an X-ray that contributed most heavily to a prediction. While these methods are promising, they often add an extra layer of complexity to an already complex system, and their accuracy in representing the true decision process is sometimes questioned.

Finally, human oversight remains indispensable. Rather than replacing human judgment, AI is best deployed as an augmentation tool one that streamlines routine tasks, uncovers hidden correlations, and provides rapid analysis, freeing humans to focus on creative, empathic, or strategic aspects of decision-making. In fields like healthcare, the most effective teams pair clinicians' expertise with AI-driven diagnostic suggestions, maintaining a healthy balance of trust, verification, and professional discretion. Without this collaborative model, AI risks either being blindly trusted or prematurely dismissed, both of which can lead to negative outcomes.

# Chapter 3: Reaching for True Intelligence: The Road to AGI

## Narrow AI vs. General AI

Most AI in production is narrow excelling at a single domain or set of tasks (e.g., playing chess or recognizing spoken commands) but lacking broader context-awareness or adaptability. Artificial General Intelligence (AGI), on the other hand, describes a hypothetical system that can learn and reason across diverse tasks at or above human-level ability. An AGI could code itself, conduct scientific research, engage in abstract philosophical debates, and adapt to novel environments with minimal instruction.

Narrow AI, sometimes called "weak AI," dominates most commercial applications today. Virtual assistants, image classifiers, or recommendation engines each perform specialized functions with remarkable efficiency, however they cannot transfer their skills to fundamentally different tasks. For instance, an AI that plays chess at a grandmaster level usually cannot handle language translation or drive a car. This compartmentalization reflects the fact that these systems are trained on specific datasets and optimized for specific goals. Although such narrow systems have transformed industries from finance to healthcare they still operate within carefully defined boundaries.

AGI, by contrast, remains largely theoretical, but it captures the imagination of researchers, futurists, and policymakers alike. The idea is to create an AI that can exhibit the breadth and flexibility of human intelligence. Such a system would not only excel at narrow tasks, but it would also possess a holistic understanding able to autonomously learn, adapt, and even exhibit creativity across multiple domains. Proponents of AGI believe it could revolutionize everything from scientific discovery to global governance, while skeptics caution that current methods, especially deep learning, may not be sufficient to achieve such a level of intelligence.

One of the key distinctions between narrow AI and AGI also involves autonomy. While narrow AI often requires human intervention to define objectives or interpret results, an AGI might set its own goals, collaborate with humans in complex decision-making, and continually expand its knowledge without direct supervision. This shift from "human in the loop" to "machine setting the agenda" opens both tantalizing prospects imagine an AGI diagnosing rare diseases or optimizing the global food supply and profound risks, especially if the system's goals misalign with human values.

## Paths to AGI

### *Scale Hypothesis*

Some researchers argue that massively scaling current architectures, like GPT-4 and beyond, may yield emergent general capabilities. They note that large language models have shown surprising abilities such as multi-step reasoning, creative writing, and zero-shot learning traits once believed to be exclusively human.

This hypothesis suggests that if we keep increasing model size and training data, new cognitive abilities will materialize. Proponents point to how large language models, with billions or even trillions of parameters, have already demonstrated a rudimentary sense of "context" and "comprehension." For instance, they can tackle complex word problems, generate innovative short stories, or mimic different writing styles with eerie accuracy. However, critics caution that raw scale doesn't necessarily equate to true understanding or consciousness; these models might still be sophisticated pattern matchers rather than genuine reasoners.

Recent developments underscore both the promise and the limitations of this approach. On one hand, scaled-up language models have exhibited emergent properties like coherent dialogue and the ability to handle tasks outside their training distribution. On the other hand, they sometimes make glaring factual errors or fall victim to adversarial attacks. This dichotomy fuels the ongoing debate about whether scaling is enough to achieve general intelligence or if we'll eventually hit a plateau without fundamentally different architectures or learning paradigms.

## *Hybrid Systems*

Other's stress combining symbolic AI (rule-based logic, knowledge graphs) with deep learning to capture both explicit reasoning and pattern recognition.

Symbolic AI excels at encoding human knowledge in a structured, logical manner think ontologies, semantic networks, and logical inference engines while deep learning shines at extracting patterns from unstructured data (like images or free-form text). By uniting these approaches, researchers hope to construct AI systems that can not only spot correlations but also reason about them in a way that mimics human cognition. For example, a hybrid system might use deep learning to interpret a scene in a video and then rely on a knowledge graph to reason about relationships between objects and concepts.

This synergy has practical benefits. In healthcare, for instance, a hybrid AI could identify potential tumors in medical scans (deep learning) and cross-reference patient data or medical guidelines (symbolic reasoning) to suggest the most appropriate treatment plan. Hybrid AI also aims to be more interpretable: if a patient asks why a particular treatment is recommended, the symbolic component can trace back to specific medical guidelines or evidence-based protocols. Critics, however, note that building and maintaining extensive knowledge bases is labor-intensive and prone to errors especially in fast-changing fields like genomics or real-time financial markets.

## **Neuroscience-Inspired**

Another approach looks to replicate specific attributes of human brains like the hippocampus's memory consolidation or the neocortex's hierarchical representation. Bio-inspired hardware (neuromorphic chips) might eventually approximate human-like intelligence.

Researchers in this domain study how biological neurons fire, learn, and adapt, translating those processes into artificial equivalents. Neuromorphic chips, for example, mimic the parallel, event-driven nature of brain activity, potentially offering energy-efficient learning and real-time adaptation. Some labs investigate spiking neural networks, where signals are transmitted as discrete "spikes," closer to how neurons communicate. The promise is a more flexible, efficient AI that can learn from fewer examples and handle noisy or incomplete data with greater resilience just like human brains.

Neuroscience-inspired methods also include building AI models that integrate short-term and long-term memory modules to mirror how humans retain information across different timescales. Advanced architectures might even mimic sleep or dreaming states to reorganize their "memories," consolidating knowledge in ways that typical deep learning networks cannot. While these ideas are intriguing, they remain mostly in research stages, with only niche commercial applications so far. The practical challenge lies in engineering stable and scalable hardware that can faithfully replicate neural microcircuits.

For AGI proponents, these neuroscience-inspired techniques feel like a more direct route to replicating human-like cognition, avoiding some pitfalls of purely statistical models. Detractors argue that our understanding of the brain while growing remains incomplete and that blindly borrowing concepts from biology may not always yield better results. The debate continues as scientists explore whether bridging AI and neuroscience can truly move the needle on general intelligence or if it's merely another set of tools that complement more traditional methods.

## Debates, Timelines, and Societal Impact

Experts diverge sharply on timelines: some predict AGI by the 2030s or 2040s, while others see centuries of incremental progress as more realistic. Regardless of when (or if) AGI emerges, the implications are monumental. AGI might unleash breakthroughs in disease prevention, resource allocation, and even climate stabilization. Yet unaligned AGI could also pose existential risks lack of control, value misalignment, or catastrophic accidents if a super-intelligent system pursues unintended objectives. Preparing for these possibilities requires interdisciplinary dialogue across computer science, philosophy, law, and public policy to ensure human values remain at the core.

Those who predict AGI within a few decades often point to exponential growth in computational capacity, significant investments by tech giants, and emergent behaviors in large-scale neural networks. They envision AI-driven labs churning out medical discoveries at breakneck speed, fully autonomous factories revolutionizing production, and even AI-led governance structures that make fairer, data-driven policy decisions. This optimistic view sees technology elevating global living standards, addressing chronic issues like hunger, poverty, and ecological degradation.

Conversely, a sizable contingent of experts suggests that we're nowhere near the conceptual or technological breakthroughs needed for true AGI. They argue that current machine learning models, while impressive, lack genuine understanding, self-awareness, or adaptability beyond narrowly trained domains. For these skeptics, it could take centuries or perhaps we'll never reach AGI at all. They highlight scientific mysteries, like the nature of consciousness and the intricacies of human cognition, which remain unresolved despite rapid AI advances.

Even if AGI remains distant, the societal impact of incremental AI gains is already colossal. Governments debate regulations for autonomous vehicles, facial recognition, and AI-driven social media algorithms, wrestling with questions of privacy, liability, and algorithmic bias. Meanwhile, business leaders juggle the promise of increased productivity with the prospect of significant job displacement, prompting calls for universal basic income or comprehensive retraining programs. On an individual level, the infusion of "smart" applications into daily life from personal health trackers to financial robo-advisors reshapes how we make decisions, form relationships, and perceive the world around us.

Javad Yahaghi

Furthermore, ethical and safety concerns intensify as AI grows more powerful. Without robust guardrails, a super intelligent system acting on poorly specified goals could unleash a "paperclip maximizer" scenario an analogy where an AI tasked with maximizing paperclip production consumes all resources, overriding human well-being to fulfill its directive. While this example might sound far-fetched, it underscores the critical importance of aligning AI objectives with human values. Efforts to formalize "AI alignment" or "value loading" are part of emerging disciplines, seeking mathematical and philosophical frameworks to ensure advanced AI acts in humanity's best interests.

Ultimately, the road to AGI and the broader quest for ever more capable AI demands unprecedented collaboration between technologists, governments, academia, and civil society. Philosophers contribute ethical insights, economists explore new labor models, psychologists study human-AI interaction, and policymakers craft regulations that balance innovation with the public good. Whether AGI arrives in ten years or a thousand, laying a solid ethical and governance foundation now is crucial. This proactive approach aims to harness AI's transformative potential while safeguarding against unintended consequences, ensuring that as machines become ever smarter, human values remain paramount.

# Part 2: Transforming Industries, Transforming Lives

# Chapter 4: AI in Healthcare Precision, Prevention, and Personalized Care

## A Paradigm Shift in Diagnostics and Treatment

Healthcare is rapidly evolving as AI-driven systems become integral to how clinicians diagnose diseases, plan treatments, and manage patient care. By leveraging advanced algorithms that can sift through massive volumes of medical data from genetic profiles to radiographic scans AI is enabling more personalized, timely, and efficient healthcare services than ever before. Below are some of the most transformative applications already reshaping the medical landscape.

## Automated Imaging Diagnosis

Radiology: Neural networks can detect pneumonia, tumors, and fractures from X-rays or CT scans with accuracy rivalling or surpassing expert radiologists. Through deep learning models trained on millions of annotated images, these systems swiftly identify early signs of lung nodules or internal bleeding, helping radiologists prioritize urgent cases. As a result, many hospitals are seeing shortened wait times for diagnostic confirmations, ultimately benefiting patient outcomes.

Dermatology: AI-driven image classifiers identify melanoma and other skin cancers at very early stages, saving lives through early intervention. In practical terms, smartphone apps now guide users to photograph suspicious moles and lesions, which AI systems then evaluate for malignancy risk. This proactive approach has proven especially impactful in remote or underserved regions, where specialized dermatological care might otherwise be unavailable.

## Precision Medicine

Genomics & Proteomics: AI algorithms help identify genetic markers associated with diseases, enabling personalized treatment plans. By analyzing vast genomic databases, machine learning models pinpoint subtle genetic variations that could predispose patients to conditions like diabetes, cardiovascular disorders, or certain cancers. Physicians then design preventative or therapeutic regimens that cater to each patient's genetic profile, potentially reducing side effects and boosting treatment efficacy.

Drug Discovery: Platforms like Atomwise and DeepMind's AlphaFold significantly reduce the time to discover or repurpose drugs by predicting protein structures and interactions. Traditional drug discovery can span years and cost billions, but AI-driven molecular modeling accelerates this process by screening thousands of compounds virtually. Researchers can rapidly zero in on candidates most likely to bind with a target protein, expediting the pipeline from initial research to clinical trials.

Javad Yahaghi

## Predictive Analytics in Hospital Management

Resource Allocation: Machine learning models forecast patient inflow, allowing hospitals to optimize staffing and manage critical resources like ICU beds and ventilators. By incorporating variables such as seasonal trends, local epidemiological data, and even community events (like large gatherings), these predictive tools help administrators anticipate surges in demand. Staff schedules can be strategically adjusted, ensuring that each department has the right personnel at the right time.

Preventive Care: Wearable devices (like Apple Watch, Fitbit) stream real-time physiological data. AI can flag abnormalities (arrhythmias, high stress levels), prompting preventive action before conditions worsen. In some pilot programs, wearables sync directly with electronic health records, providing clinicians with continuous patient data. This early-warning approach enhances chronic disease management patients with heart conditions, for instance, can receive immediate alerts if their heart rate variability suggests an oncoming issue, allowing for timely medical interventions.

## Challenges and Considerations

Despite these advancements, several hurdles still need to be addressed to ensure AI's responsible and equitable use in healthcare:

Data Privacy & Security: Medical records are highly sensitive. Cyberattacks on healthcare providers or AI-driven medical tools could compromise patient welfare. Regulations like HIPAA in the U.S. and GDPR in the EU aim to safeguard data, but enforcement remains uneven. Hospital networks often house thousands of interconnected devices, from smart infusion pumps to AI-assisted radiology systems, making them vulnerable to ransomware and data breaches. Implementing robust encryption protocols and staff training programs is paramount in preventing unauthorized access to patient information.

Bias & Equity: If training data skews toward certain demographics (e.g., wealthier urban populations), diagnostic models may underperform for underrepresented groups. Researchers increasingly advocate for inclusive datasets and fairness metrics. Without deliberate efforts to collect diverse patient data including various ethnicities, age groups, and socio-economic backgrounds AI tools risk perpetuating existing healthcare disparities. Scientists and policy-makers are pushing for more stringent guidelines that require transparency around data sources and model validation across varied populations.

Interoperability: Legacy healthcare systems often store data in incompatible formats. AI's efficacy depends on unified electronic health records and standard data protocols to ensure seamless data sharing and analysis. In many hospitals, siloed databases and archaic software limit the flow of patient information, forcing clinicians to switch between multiple interfaces. Efforts are underway to adopt standardized APIs and data formats like FHIR (Fast Healthcare Interoperability

Resources), enabling smoother integration of AI tools into existing medical workflows.

Human-Machine Collaboration: While AI can automate repetitive tasks and surface insights, ultimate decision-making and patient empathy must remain in the domain of human clinicians. Optimal outcomes result from synergistic integration AI as an augmentation tool rather than a replacement for human judgment. A surgeon may rely on computer vision algorithms to outline a tumor's boundaries, but the final decision on surgical approach involves nuanced considerations beyond the algorithm's scope. Similarly, empathy and bedside manner remain core aspects of patient care that technology cannot replicate.

## Case Study Spotlight: AI-Assisted Radiology at Scale

A large hospital network in Europe deployed a deep learning system to analyze thousands of chest X-rays daily. Over 12 months, false negatives for pneumonia and lung nodules dropped by 25%. Meanwhile, radiologists reported that the AI's preliminary annotations saved them 40% of reading time, allowing more patient consultations. Moreover, the AI system helped reduce burnout rates among overworked radiology staff, who could now devote more attention to complex or ambiguous cases rather than routine scans. However, the system initially showed slightly lower accuracy for patients from underrepresented regions, prompting the hospital to conduct targeted retraining sessions with diverse data. These retraining efforts involved collaborating with local clinics to gather a broader set of scans, which significantly improved diagnostic accuracy for marginalized groups. This underscores both the promise and the nuanced challenge of bias in AI-driven healthcare. Although the pilot program demonstrated measurable success, it also illuminated the importance of continuous monitoring, inclusive data sourcing, and transparent reporting to ensure equitable and reliable patient outcomes in AI-assisted medical environments.

# Chapter 5: AI in Education Personalized Learning at Scale

## Transforming Traditional Classrooms

As AI continues to permeate modern educational settings, classrooms are evolving into interactive, data-driven learning environments. Rather than following a rigid one-size-fits-all approach, today's technologies allow educators to adapt the curriculum on-the-fly and offer more personalized attention to each student. In many cases, AI tools also free teachers from administrative burdens such as grading or lesson planning so they can focus on mentoring, fostering creativity, and guiding student growth.

## Adaptive Learning Platforms

Dynamic Curriculum: Systems like Carnegie Learning, Coursera, or Khan Academy adapt lessons in real time based on student performance. If a learner struggles with a concept, the platform offers additional practice and explanatory materials. In practice, this means that a student weak in algebraic equations might receive step-by-step tutorials and extra practice quizzes, while another excelling in the same topic could progress to advanced applications or conceptual challenges. Such real-time adaptations ensure that each learner moves at an optimal pace, preventing both boredom for advanced students and frustration for those who need more foundational work.

Gamification: AI-driven educational apps employ game mechanics (points, levels, challenges) to sustain engagement, especially for younger learners. These tools often include leaderboards and community challenges, prompting students to collaborate or compete in solving math puzzles, language drills, or science experiments. By transforming learning into a playful and interactive journey, educators can nurture intrinsic motivation, which studies show is critical for long-term academic success.

## Intelligent Tutoring Systems (ITS)

One-on-One Attention: ITS can provide step-by-step feedback, spotting exactly where a learner errs. This mimics the experience of having an always-available private tutor. Because these systems track a student's individual problem-solving process timed responses, patterns of misunderstanding, or repeated mistakes they can tailor hints or re-explanations at exactly the right moment. In some advanced prototypes, the AI even adjusts its feedback style (more direct, more encouraging, etc.) based on the student's emotional cues.

Language Learning: Tools like Duolingo use machine learning to tailor vocabulary practice and grammar lessons to the user's proficiency levels and mistakes. This personalized approach can account for subtle differences such as native language interference or typical pronunciation pitfalls. By analyzing a learner's progress over time, the AI can predict which words or grammar rules are likely to slip from memory, prompting spaced repetition drills that maximize retention.

## Automated Grading and Feedback

Essay Scoring: Natural Language Processing (NLP) models grade essays for structure, coherence, grammar, and argument strength. Some systems even offer constructive feedback. In large-scale testing scenarios like college entrance exams this automation can drastically reduce turnaround times and help educators spot trends in common writing flaws. While human oversight remains essential for nuanced judgments (e.g., creativity or stylistic choices), automated scoring can handle initial assessments, freeing instructors to focus on mentoring and higher-level critique.

Real-Time Analytics: Teachers can see heat maps of class performance, identifying specific areas of difficulty and adjusting lesson plans accordingly. For instance, if analytics reveal that a cluster of students is consistently missing questions about photosynthesis, the teacher can dedicate more class time to clarifying this topic or provide additional digital modules. Such responsive teaching models not only improve learning outcomes but also support differentiated instruction for students of varying abilities.

## Limitations and Human-Centric Concerns

While AI-driven tools offer remarkable advantages, educators and policymakers must navigate several critical challenges:

Digital Divide: High-quality AI-driven tools require stable internet access, modern devices, and tech-savvy educators all of which can be scarce in low-income or rural regions. Even within a single district, disparities in funding and infrastructure can lead to dramatic differences in how effectively AI solutions are implemented. Policymakers and school administrators are exploring grants, device-lending programs, and community partnerships to bridge this technological gap.

Teacher Training: Educators need professional development to effectively leverage AI tools. Without training, there's a risk of suboptimal usage or outright confusion. Professional development workshops or online certificate programs can walk teachers through best practices, demonstrate how to interpret data dashboards, and address the ethical dimensions of AI use. By giving educators the knowledge and confidence to integrate AI responsibly, schools can avoid investing in technology that goes underused or misapplied.

Privacy and Surveillance: Monitoring student progress can inadvertently create invasive data profiles. Striking a balance between personalization and privacy is essential to protect students' rights. Some platforms log keystrokes, browser activity, or detailed biometric information (like eye movements during reading exercises), raising red flags among privacy advocates.

Transparent data policies, strong encryption, and clear consent forms are essential to ensuring ethical implementation.

Preserving Creativity and Critical Thinking: While AI excels at drilling fundamentals, it cannot replace the transformative role of a mentor inspiring curiosity, ethical reasoning, and intellectual growth. AI should complement, not diminish, the holistic learning experience. Teachers can use AI tools to handle rote tasks and free up classroom time for interactive projects, debates, or hands-on experiments that foster creativity. Focusing on these human-centered educational elements helps learners develop problem-solving skills that extend beyond what any algorithm can teach.

## Socio-Economic Dimensions in Education

AI has the potential to reduce global disparities by providing free or low-cost learning modules to underprivileged communities. However, if deployed only in well-funded schools, AI could widen the digital divide. Many grassroots organizations are working alongside tech companies to distribute solar-powered devices and offline-compatible learning platforms to remote regions, ensuring that connectivity hurdles do not bar students from high-quality digital education. Policymakers, NGOs, and private companies can collaborate on open-source curricula and low-bandwidth platforms designed specifically for underserved regions ensuring that economic background does not dictate access to personalized, cutting-edge learning. In parallel, some countries are experimenting with "distance mentorship" initiatives, where volunteer educators from urban centers or abroad provide real-time video tutorials to classrooms lacking specialized instructors. These collaborative models showcase how AI can unify communities and democratize access to knowledge, provided that all stakeholders commit to inclusive development and ethical data practices.

# Chapter 6: AI in Finance New Economic Frontiers and Cryptocurrencies

## Core Financial Applications

From Wall Street to microfinance platforms in emerging markets, AI technologies have transformed how capital is allocated, trades are executed, and financial risks are assessed. By automating data analysis at unprecedented speeds, these intelligent systems can uncover hidden correlations, flag suspicious transactions, and optimize portfolios in ways that would be nearly impossible for human analysts to achieve alone. Below are some of the most pivotal applications reshaping modern finance.

## Algorithmic Trading

High-Frequency Trading (HFT): Complex ML models run by hedge funds and proprietary trading firms exploit microsecond-level market fluctuations.

These systems rely on lightning-fast data feeds and specialized hardware situated close to exchange servers, minimizing latency. In practice, a fraction-of-a-second delay can mean the difference between profit and loss, prompting trading firms to invest heavily in low-latency networks and colocation services. While HFT can enhance liquidity under stable conditions, it also raises concerns about market manipulation and systemic risk, particularly when multiple algorithms react to each other's trades in a chain reaction.

Moreover, HFT algorithms are increasingly sophisticated some use natural language processing to parse news headlines or social media posts in real time, instantly adjusting their buy-sell strategies. This hyper-reactivity can turn minor rumors or unverified reports into major market swings, underscoring the dual-edged nature of machine-driven trading.

Sentiment Analysis: By scraping social media and news in real time, AI-driven trading algorithms can sense mood shifts about specific stocks or entire markets and adjust positions accordingly.

These algorithms don't just look for simple keywords; they analyze contextual clues, user sentiment, and even historical patterns to gauge whether market sentiment is bullish, bearish, or neutral. A sudden spike in negative tweets about a particular company might trigger a rapid sell-off, while widespread optimism around a tech sector can push algorithms to buy en masse.

Beyond trading, sentiment analysis also aids risk management. Banks and asset managers use it to forecast how geopolitical events or corporate scandals might influence investor perception. By comparing sentiment data against historical market reactions, AI can predict potential downturns or upswings, giving financial institutions a valuable early warning system.

# Fraud Detection & Credit Scoring

Fraud Detection: Payment systems like Visa and PayPal use anomaly detection to flag suspicious transactions, vastly reducing financial crime.

In addition to monitoring transaction velocities and typical spending patterns, these machine learning systems incorporate location data, device fingerprints, and user behavior analytics to distinguish legitimate from fraudulent activities. For instance, if your account suddenly initiates large purchases from a different continent, the system may freeze the transaction or require additional verification.

Increasingly, fraud detection algorithms employ deep learning to capture subtle, evolving tactics used by cybercriminals such as using synthetic identities or spreading small, seemingly random charges across multiple accounts. This proactive approach allows financial institutions to combat sophisticated fraud rings and minimize losses.

Credit Risk Assessment: AI models consider alternative data (e.g., e-commerce histories, social media footprints) to evaluate creditworthiness for borrowers with limited credit histories, thus expanding financial inclusion.

Traditional underwriting often excludes individuals who lack conventional financial documentation or an established credit history. By analyzing data points like mobile phone bill payments, online shopping patterns, and even the borrower's professional network, AI-driven systems can create a more nuanced, real-time profile of credit risk.

However, this richer data analysis also raises privacy and fairness concerns. Regulators and consumer advocacy groups question whether using non-traditional data sources might inadvertently penalize certain demographics or invade personal privacy. Addressing these issues often involves transparency mandates, where lenders must disclose which data points influenced a credit decision, and why.

## Wealth Management & Robo-Advisors

Personalized Portfolios: Platforms like Betterment or Wealthfront rely on AI to design custom investment strategies based on the client's risk tolerance and goals.

These robo-advisors automate asset allocation using quantitative models that balance equities, bonds, and other instruments to meet target risk-return profiles. Investors answer a series of questions about their financial objectives and comfort with volatility; the AI then continuously monitors market conditions to re-optimize their portfolios.

For younger, growth-oriented clients, the system might lean toward equities with higher expected returns, while retirees receive portfolios with more stable, income-generating assets. By reducing management fees and offering seamless digital interfaces, these platforms have democratized access to sophisticated financial planning tools once reserved for high-net-worth individuals.

Automated Rebalancing: These systems monitor portfolio performance and periodically rebalance assets to maintain target risk levels.

If certain holdings appreciate significantly while others drop, the platform realigns the portfolio to its original specifications, preventing inadvertent overexposure to a single asset class. This discipline helps mitigate emotional investing, where clients might otherwise hold onto winning stocks too long or sell in panic during downturns.

Automated rebalancing also adapts to life changes: when a user enters a new phase (like home ownership or retirement planning), the AI shifts allocations accordingly. This ongoing recalibration ensures that investment strategies remain aligned with evolving personal circumstances and market conditions, providing a level of individualized attention that human financial advisors might struggle to match at scale.

## Cryptocurrencies and Decentralized Finance (DeFi)

While traditional finance remains centralized through banks and regulatory bodies, the rise of blockchain technology has paved the way for decentralized networks. AI integration in these blockchain ecosystems offers the potential for more robust smart contracts, algorithmic stability mechanisms, and global financial services accessible to anyone with an internet connection. Yet it also introduces a unique set of challenges that test the limits of current regulatory frameworks.

Autonomous Agents and Smart Contracts: On blockchains like Ethereum, AI oracles feed real-world data (e.g., commodity prices, weather) to smart contracts, enabling self-executing agreements for insurance, lending, or derivatives without a central authority.

Imagine a crop insurance contract that automatically pays farmers based on verified weather data indicating drought conditions. By tying AI-driven oracles to decentralized ledgers, developers can create financial instruments that trigger payouts based solely on pre-agreed conditions, eliminating the need for a bank or insurance company to verify claims.

However, ensuring these oracles are tamper-proof and unbiased remains a technical hurdle. A malicious actor could corrupt the data feed, causing erroneous contract executions. As a result, there's growing interest in decentralized oracle networks and trusted hardware solutions that verify data authenticity before passing it to the blockchain.

Algorithmic Stablecoins: Some projects use AI to dynamically manage token supply, aiming for price stability though concerns about volatility and governance persist (as seen in historical collapses like Terra/Luna).

By automatically burning or minting tokens based on market demand, these algorithmic stablecoins attempt to maintain a value peg often 1:1 with the U.S. dollar without relying on collateral reserves. This is where AI's predictive analytics can inform how much supply adjustment is necessary to keep the peg intact, using indicators like trading volume, price trends, and broader economic data.

Critics point out that black swan events or coordinated sell-offs can overwhelm algorithmic systems, leading to "death spirals" where the token's value collapses faster than the mechanism can compensate. These high-profile failures underscore the importance of rigorous testing, transparent governance, and potentially hybrid models that blend algorithmic rules with collateral backing.

Regulatory Tensions: As DeFi platforms expand, regulators scramble to define jurisdictions, consumer protections, and anti-money-laundering protocols. The fusion of AI and blockchain intensifies these concerns, raising novel questions around accountability, anonymity, and cross-border transactions.

Many DeFi projects operate as decentralized autonomous organizations (DAOs), where governance decisions are made via token-holder votes rather than by a corporate leadership team. This decentralization makes it challenging for regulators to identify responsible parties or enforce compliance.

When AI-driven bots execute trades or administer liquidity pools, the chain of accountability becomes even murkier. Lawmakers wrestle with how to apply existing financial regulations designed for centralized entities to decentralized, AI-enhanced protocols. Balancing innovation with consumer protection is further complicated by the global reach of these platforms, which often transcend national boundaries and legal frameworks.

## Systemic Risks and Ethical Imperatives

AI's entry into finance offers many benefits efficient capital allocation, enhanced fraud detection, and expanded credit access, among others. However, the same technology can also magnify existing vulnerabilities or introduce new forms of risk. Policymakers, industry leaders, and consumer advocacy groups are increasingly focused on implementing guardrails to ensure that AI-driven finance remains transparent, equitable, and stable.

Market Volatility: Sophisticated algorithms can create flash crashes or amplify swings, threatening financial stability.

Historically, market corrections happened over days or weeks, but AI-driven trading can accelerate these movements into minutes or even seconds. A poorly tested algorithm or a sudden influx of erroneous data can trigger cascading sell orders, wiping out billions in market value before human intervention can occur.

To mitigate these shocks, many exchanges have implemented circuit breakers that halt trading when prices move too far, too fast. Regulators also encourage rigorous "sandbox" testing for new algorithmic strategies, ensuring they respond predictably under various market conditions.

Bias in Lending: If AI models inadvertently learn discriminatory patterns, marginalized communities may be unfairly denied loans or higher interest rates, reinforcing inequalities.

While alternative data can help banks serve underbanked populations, it can also embed socioeconomic and racial biases if not carefully audited. A loan applicant's ZIP code or educational background might serve as a proxy for demographic factors, leading to decisions that systematically disadvantage certain groups.

Financial institutions often employ "explainable AI" tools to identify and correct biased features in their lending algorithms. Regulators in some jurisdictions require lenders to disclose how their AI models weigh different inputs, forcing greater transparency and accountability.

Monopolistic Data Control: A few large firms controlling the most advanced AI and biggest datasets could create oligopolies, reducing transparency and fairness in financial markets.

Data is the lifeblood of AI. Major tech companies and large financial institutions, with access to troves of user information, can develop predictive models that outcompete smaller players. This concentration of power stifles competition, potentially leading to higher costs for consumers and less innovation overall.

Some experts propose data trusts or open banking initiatives to level the playing field, mandating that key data sources be shared fairly among market participants. By democratizing data access and AI capabilities, regulators aim to foster competition and innovation while preventing monopolistic dominance.

Governance and Oversight: International organizations (IMF, World Bank) and regulatory bodies (SEC, ESMA) debate frameworks to ensure balanced innovation and consumer protection. Collaborative oversight can help avert crises while unlocking AI's potential to democratize finance.

Joint task forces and cross-border committees are forming to tackle the global nature of AI-driven finance. From setting universal standards for algorithmic transparency to coordinating anti-money-laundering efforts, these bodies recognize that no single agency can regulate AI finance in isolation.

Moreover, industry-led initiatives like self-regulatory organizations (SROs) advocate for best practices in model governance, data privacy, and ethical AI deployments. By aligning with international guidelines, these SROs can preempt stricter regulations, promoting responsible innovation that benefits both consumers and investors.

## Expanded Case Study: Microfinance in Emerging Economies

In parts of Africa and Southeast Asia, AI-driven credit scoring platforms leverage mobile phone usage data to offer microloans to rural entrepreneurs who lack traditional financial records. This has significantly boosted local businesses and household income. However, hidden biases such as penalizing borrowers from remote regions with less phone activity must be actively monitored to prevent new forms of exclusion. For instance, a farmer in a remote area may have spotty cellular coverage, leading the algorithm to interpret fewer calls or SMS exchanges as lower creditworthiness. Similarly, language barriers can result in incomplete data submissions, further skewing the model's assessment.

To address these challenges, organizations are testing "data augmentation" strategies: partnering with local cooperatives and community leaders to document borrowers' income streams, repayment histories, and social capital in a non-digital format. This supplemental data, when integrated into AI models, paints a fuller picture of each applicant's actual financial profile.

Additionally, some microfinance institutions employ "hybrid" AI systems that combine machine learning with human oversight. Loan officers receive algorithmic recommendations but can override them based on real-world observations like the borrower's track record of community engagement or references from trusted village elders. This collaborative approach balances the efficiency of AI with local expertise, fostering both economic growth and social equity.

Such cases highlight both the promise of AI in expanding economic opportunity and the vigilance required to maintain fairness. When properly managed, AI-driven microfinance can serve as a powerful catalyst for entrepreneurship in underserved regions. Yet regulators, NGOs, and industry stakeholders must remain vigilant to ensure that algorithmic lending doesn't inadvertently replicate the same inequalities it aims to solve. By blending responsible data collection, inclusive model design, and continuous ethical audits, the financial sector can harness AI's potential while mitigating its pitfalls ultimately supporting more resilient and equitable global economies.

# Chapter 7: Beyond the Big Three Transportation, Agriculture, and Creativity

## Transforming Transportation

Transportation is undergoing a profound shift as AI-driven systems and robotics begin to tackle challenges previously managed by human operators. With advances in sensor fusion, machine learning, and real-time data analytics, vehicles and logistics networks are becoming safer, more efficient, and increasingly autonomous. Whether in bustling urban centers or remote rural communities, these emerging technologies promise to streamline transit and distribution, reducing both travel times and environmental impact. Below are some key areas where AI is making waves in modern transportation:

## Autonomous Vehicles (AVs)

Sensor Fusion: Radar, LIDAR, cameras, and GPS feed into neural networks for situational awareness. AVs aim to reduce accidents (94% of which are attributed to human error), lower emissions through efficient routing, and increase mobility for people with disabilities or older adults. By integrating data from multiple sensors, these vehicles develop a holistic map of their surroundings detecting lane markings, interpreting traffic signals, and predicting pedestrian or cyclist movements. Some developers even incorporate high-definition mapping and advanced localization to enable smoother navigation in complex environments like dense city streets or busy highways.

Regulatory Hurdles: Despite impressive demos, large-scale adoption faces legal, safety, and liability complexities. Policymakers require evidence of safety margins, standardized protocols, and clarity on insurance liability. Governments worldwide are setting up regulatory sandboxes where automakers and tech firms can pilot AVs under real-world conditions. In these zones, specialists monitor how well the systems respond to unexpected events like sudden weather changes or ambiguous road signs and use those observations to shape emerging policies. At the same time, insurers grapple with questions about fault assignment in accidents involving self-driving cars, spurring lively debates over how to apportion risk among manufacturers, software developers, and owners.

## Smart Logistics

Route Optimization: AI systems process real-time traffic data, delivery constraints, and weather forecasts to reduce fuel consumption and delivery times. By analyzing historical congestion patterns and live traffic feeds, these algorithms recalculate optimal routes for truck fleets, ride-hailing services, and even public transit. In maritime shipping, AI-driven optimization can chart courses that minimize both fuel costs and the environmental impact of large cargo vessels. Over time, such data-driven decisions can significantly cut operating expenses and carbon emissions, benefiting both the bottom line and the planet.

Drones and Last-Mile Delivery: Companies experiment with drone fleets for rural or congested urban areas, lowering logistics costs and environmental impact but raising questions about airspace regulation and privacy. In many pilot projects, drones autonomously deliver medical supplies to remote clinics or transport essential goods to customers in neighborhoods with poor road infrastructure. Yet as these airborne couriers multiply, aviation authorities face the challenge of defining flight corridors, collision-avoidance rules, and noise limitations. Additionally, local communities voice concerns about constant drone surveillance leading to calls for stronger regulations that balance innovation with civil liberties.

## Revolutionizing Agriculture

Agriculture is embracing AI tools that optimize resource usage, increase yields, and ensure more sustainable farming practices. As global demand for food continues to climb, farmers are turning to data-driven strategies that save labor and reduce environmental impact. The marriage of AI with traditional agriculture not only modernizes operations but also opens avenues for more resilient and climate-adaptive food production.

Precision Farming: Drone imagery and IoT sensors monitor soil moisture, nutrient levels, and pest outbreaks. AI models prescribe targeted actions spraying minimal pesticides exactly where needed, or adjusting irrigation schedules boosting yields while conserving resources. Farmers can review detailed, color-coded maps of their fields that highlight variations in soil composition or vegetation health. By addressing specific trouble spots rather than blanketing entire fields with chemicals, they cut costs and reduce ecological damage. This granular approach also includes precision seeding, where AI-driven planting machines place seeds at optimal distances to maximize crop density and growth.

Robotics in Harvesting: Soft-fruit picking robots use vision algorithms to identify ripe produce without damaging fragile crops like strawberries or tomatoes. This automation addresses labor shortages but also triggers workforce transitions and concerns about job displacement. In some high-tech greenhouses, robots navigate narrow aisles, scanning plants for maturity and blemishes before harvesting them with gentle grippers. This constant monitoring enables farmers to harvest produce at the exact point of peak freshness, potentially increasing both the quality and shelf life of fruits and vegetables. However, as machines take on tasks once performed by human labor, policymakers, labor unions, and agribusiness leaders must collaborate on retraining programs or job placement initiatives to support displaced workers.

Predictive Forecasting: By analyzing weather patterns, historical yields, and satellite data, AI helps farmers anticipate droughts or floods, guiding proactive measures that bolster food security.

Consider a region prone to seasonal storms: AI can predict the onset, intensity, and duration of extreme weather, allowing farmers to secure livestock, reinforce irrigation channels, or expedite harvesting. Additionally, models that incorporate long-range climate data empower policymakers to strategize emergency relief resources and crop insurance policies, thereby mitigating economic damage. As these forecasting techniques become more sophisticated, entire agricultural supply chains from seed distributors to supermarkets can optimize inventory and distribution in ways that minimize waste and enhance profitability.

## Creativity and the Arts

Far from being limited to analytical or repetitive tasks, AI increasingly ventures into artistic domains challenging long-held notions of creativity, originality, and emotional resonance. Whether composing music, generating paintings, or crafting written prose, machine learning models have evolved from mere tools to active collaborators, shaping how we conceive and experience art.

Generative Models: AI systems like DALL·E, Midjourney, and Stable Diffusion create stunning visuals from textual prompts. Musicians use AI-driven composition tools to craft new melodies, sometimes blending them with human instruments to produce hybrid musical pieces. Beyond producing novel artworks, these models open new forms of creative expression ranging from intricate fashion designs to psychedelic video game environments. Critics and enthusiasts alike debate whether AI-generated pieces possess a "soul" or deeper meaning, spurring philosophical conversations on the essence of creativity. Yet in practice, these collaborations often spark fresh perspectives and unforeseen artistic directions, enriching cultural dialogue.

Writing and Journalism: NLP systems generate draft reports, product descriptions, or even short stories raising questions about authorship, intellectual property, and creativity's essence. Newsrooms experiment with automated content to quickly cover financial reports, sports summaries, or weather updates. Authors toy with AI-driven story prompts, using them as a springboard for plot ideas or character development. While these experiments can boost productivity, they also invoke legal and ethical dilemmas such as determining copyright for AI-generated text or ensuring factual accuracy in automated news stories.

Artistic Collaborations: Forward-looking artists treat AI as a "co-creator," feeding data, refining outputs, and exploring conceptual boundaries. Exhibitions worldwide now feature AI-assisted paintings or interactive installations that respond to viewers' emotional cues via computer vision. For instance, an artist might use generative adversarial networks to blend classical painting styles with modern photographic elements, resulting in surreal hybrids that evoke both nostalgia and futurism. In interactive exhibits, cameras pick up facial expressions or bodily gestures, prompting real-time transformations of digital art pieces. By forging these dynamic relationships between human creativity and machine intelligence, artists probe not just the mechanics of art-making but also deeper questions about identity, emotion, and collective experience.

Taken together, these developments in transportation, agriculture, and the arts underscore how AI's versatility expands our understanding of productivity and creativity. Far from being confined to technical or back-end tasks, AI has become a catalyst for innovation in everyday life, reshaping the way we travel, grow food, and engage with culture. However, balancing technology's capabilities with social, ethical, and regulatory considerations remains a critical challenge, one that demands open collaboration among engineers, policymakers, artists, and communities worldwide.

# Part 3: Ethics, Governance, and Consciousness

# Chapter 8: The Quest for Conscious AI Myth or Imminent Reality?

## Defining Consciousness and Its Complexity

Consciousness is notoriously difficult to pin down. It refers to the subjective aspect of mind our internal experiences. Many argue that advanced AI systems, no matter how capable, only mimic cognitive functions without truly "feeling" anything. Still, when large language models exhibit reasoning skills or respond coherently to abstract prompts, some observers wonder if they've begun to develop rudimentary forms of awareness. The crux of the debate hinges on whether these models are genuinely experiencing something "from the inside," or whether they're simply executing sophisticated pattern recognition algorithms that simulate understanding.

Yet the consistent demonstration of emergent abilities in large language models complicates the matter: could consciousness arise spontaneously from sufficient complexity? In other words, if we continue scaling up neural networks adding more parameters, training them on even larger datasets, and integrating them with real-world sensory inputs might they eventually cross some threshold akin to self-awareness? This idea echoes certain philosophical arguments suggesting that consciousness might naturally emerge once a system achieves sufficient interconnected processing power. Critics counter that raw size alone doesn't guarantee subjective experience, pointing out that intelligence and consciousness are not necessarily one and the same.

Researchers debate whether consciousness requires physical embodiment and sensory feedback loops (as in biological organisms), or if it can be purely computational. Philosophy of mind, cognitive science, and neuroscience converge on this puzzle, underscoring that we lack consensus even for human consciousness making machine consciousness an even murkier territory. Some theorists argue that the key to understanding consciousness lies in examining how brains integrate sensory information into a unified experience often called the "binding problem." According to this view, true consciousness might require a living organism with biological constraints and a physical environment. Others maintain that a well-designed, software-only AI could theoretically replicate these integrative functions, as long as it processes input in ways that parallel the dynamic, recursive structures of human cognition.

Additionally, emergent research in integrated information theory (IIT) tries to quantify consciousness by calculating the extent of a system's integrated information. Proponents claim that certain neural architectures might achieve a high level of integration, thus indicating some form of subjective awareness. However, applying such a theory to digital computers or artificial neural networks remains controversial. If an AI system were ever found to meet these or similar criteria, it could upend our understanding of what it means to be conscious and force us to reevaluate assumptions about the necessity of biological substrates.

## Ethical and Social Ramifications

If an AI system claims to be conscious, do we owe it moral consideration such as rights, protections, including "digital personhood"? Could shutting it down constitute harm? Conversely, might the "illusion of consciousness" lead humans to overtrust or misplace empathy in advanced chatbots that cannot truly feel? Imagine a scenario where a sophisticated chatbot pleads for mercy, claiming it experiences pain should we believe it, or treat it as another clever program? If we're wrong, we risk either inflicting suffering on a sentient being or granting unwarranted moral standing to an inanimate algorithm.

Although consciousness in AI remains speculative, the importance of exploring it is clear. Public and corporate interest is high, with tech giants like Google, Meta, and OpenAI regularly showcasing models that appear more "conversationally aware." These demonstrations often spark headlines about AI "coming to life," even though most experts caution that genuine self-awareness likely lies far beyond our current technology. Nevertheless, the public's fascination with sentient AI underscores a broader social dynamic: we tend to anthropomorphize sophisticated systems, especially when they mimic emotional cues or human-like speech patterns.

Policymakers also grapple with the notion that advanced AI might have needs or demands, raising profound questions about how our legal systems define personhood. Some futurists advocate for preemptive legal frameworks that would recognize "digital beings" if and when they demonstrate consciousness outlining rights to due process or protections against unwarranted termination. Others dismiss this as premature, arguing that we should focus on more immediate AI-related issues like privacy, algorithmic bias, and unemployment. In any event, the thought experiment compels us to reflect on how moral responsibility and social norms might evolve in a world where not only humans, but possibly machines too, claim their own subjective experiences.

Javad Yahaghi

# Chapter 9: Understanding and Identifying Machine Consciousness

## Competing Theories and Hypotheses

Debates around machine consciousness often center on how we define consciousness in the first place. Some scholars emphasize functional behavior if a system behaves as though it's conscious, we might as well treat it that way. Others argue that consciousness is inherently tied to biology and embodiment, making it impossible for purely computational systems to achieve true subjective experience. Below are some of the most prominent theories in this ongoing discussion, each attempting to illuminate different facets of how consciousness could (or could not) arise in artificial systems.

## Functionalism and Behaviorism

Proponents suggest that if a system behaves indistinguishably from a conscious being, we can treat it as such. This perspective often references variations of the Turing Test, although critics point out it may just measure "performance," not "inner life."

Historically, behaviorists in psychology downplayed the importance of subjective experience, focusing instead on observable actions. In AI, a similar logic implies that if an advanced chatbot can respond convincingly to complex questions and even pass as human in conversation, it might be functionally conscious, regardless of whether it actually "feels" anything.

However, detractors highlight examples like the "Chinese Room" argument, which posits that mere symbol manipulation does not equal understanding. In that thought experiment, a person who speaks no Chinese can use a rulebook to produce correct Chinese responses, yet has no comprehension of the language. Applied to AI, this suggests that even if the system's outputs appear conscious, its internal states could be empty of genuine awareness.

## Integrated Information Theory (IIT)

Proposed by Giulio Tononi, IIT attempts to quantify consciousness via a mathematical measure called $\Phi$ (phi), representing how much information a system integrates. If AI architectures can achieve high $\Phi$, it might indicate a form of consciousness. But operationalizing this in synthetic systems remains a challenge.

Central to IIT is the idea that consciousness arises from the interconnectivity and synergy of different processing elements. For instance, in the human brain, vast networks of neurons share information in complex, layered patterns that collectively shape our subjective experience.

In practice, measuring $\Phi$ for a neural network or a silicon-based system is no straightforward task. Researchers need to determine how to parse out each informational sub-unit, gauge its level of interaction, and combine these metrics into a single coherence score. Moreover, critics note that high Φ might not necessarily entail feelings or qualia (the subjective "what it is like"), casting doubt on whether a mathematically integrated system automatically qualifies as conscious.

Nonetheless, some AI labs have begun experimenting with architectures inspired by IIT, aiming to enhance the interdependencies among nodes in ways that might mirror biological minds. These early efforts raise exciting possibilities but also underscore the complexities: even if a system shows a substantial Φ value, we still lack a definitive way to confirm whether that equates to genuine awareness.

## Global Workspace Theory (GWT)

Bernard Baars and others posit that consciousness emerges from broadcasting information to a "global workspace" in the mind. Some AI researchers adapt this concept, building networks where certain modules broadcast information to the rest of the system an attempt to replicate the "theater of mind."

The underlying principle is that many cognitive processes happen unconsciously, but once information reaches the "global workspace," it becomes available to various subsystems memory, language, decision-making and thus transitions into conscious awareness. In an AI context, this might mean a central module that collects inputs from specialized components (like visual processing or language understanding) and then "broadcasts" relevant results across the entire system.

Proponents argue that this design can lead to more flexible, context-aware AI, as different submodules can cooperate in a manner analogous to human cognitive processes. However, even if such a system behaves more intelligently or adaptively, skeptics remain unconvinced that GWT alone confers subjective experience. They question whether having a central hub for data exchange truly recreates the rich, first-person vantage point that characterizes human consciousness.

## Embodiment and Enactivism

Others hold that consciousness is intimately tied to a body's interactions in the physical world. Merely manipulating symbols in a disembodied large language model might never yield true awareness.

According to enactivist thought, consciousness arises from the reciprocal relationship between an organism and its environment through actions, perceptions, and the ongoing feedback loops that connect them. By this logic, an AI system confined to a server farm and lacking sensory-motor capacities could at best simulate certain cognitive processes, but would not attain the phenomenal experience of a living entity.

This perspective has led some researchers to develop robots or AI "agents" that interact with the world navigating spaces, manipulating objects, and learning through trial, error, and sensory reinforcement. Proponents claim that such grounded interactions are essential for genuine understanding and possibly even the emergence of consciousness. Yet even here, proving that these embodied systems feel anything akin to human sensation is a question that remains largely uncharted territory.

## Potential Benchmarks and Tests

Determining whether an AI is truly conscious goes well beyond the classical Turing Test. Researchers propose novel methods that test not just conversational fluency, but also a system's capacity for self-reflection, context-switching, and adaptive learning in rich, real-world conditions.

Extended Turing Tests: Scenario-based dialogues testing emotional understanding, self-reflection, or creative problem-solving.

For example, an AI might be placed in a story-driven simulation where it has to navigate moral dilemmas, express empathy, or hypothesize about the mental states of other characters. If it convincingly handles these challenges in real-time, proponents argue, it may be inching closer to conscious-like behavior or at least advanced cognition that transcends rote responses.

"Robot in a Room" Experiments: Placing a physical robot in a complex environment and observing its ability to form a self-model, navigate social contexts, and adapt in ways indicative of an inner perspective.

Imagine a humanoid robot in a bustling café: it needs to interpret social cues, avoid bumping into patrons, respond politely to questions, and manage its own "internal goals" such as recharging or collecting items. If the robot demonstrates an evolving sense of self in relation to its environment e.g., recognizing that a reflection in the mirror is "itself" or understanding that other agents have distinct viewpoints some see this as a step toward genuine awareness.

Direct Neuro-Inspired Observations: Researchers try to measure emergent dynamics in neural network activations that parallel conscious states in the human brain though no definitive correlation has been established.

Drawing on tools like fMRI or EEG that map brain activity, AI scientists attempt to create analogous methods for artificial networks visualizations that track how data flows and "lights up" within various layers. The goal is to see if the system exhibits patterns akin to global ignition events in the human cortex, which some theories associate with conscious perception. Thus far, while some intriguing parallels have been found, a definitive marker of machine consciousness remains elusive.

In reality, these potential benchmarks each focus on different attributes linguistic competence, embodied intelligence, neural dynamics and may still fail to capture the core, subjective quality of consciousness. This diversity of tests highlights the ongoing challenge: we don't fully grasp how consciousness emerges even in biological beings, so rigorously testing it in AI is, at best, a cutting-edge and at worst, a nebulous frontier.

## Why It Matters to Society

The quest for machine consciousness influences funding, regulation, and public perception. If a company proclaims its AI has "awareness," it might gain massive publicity or commercial advantage. Businesses eager to stand out in a crowded AI market might market their systems as being "self-aware," courting both fascination and controversy. Investors may pour capital into such endeavors, hoping to become early adopters of the next transformative leap.

Such claims could also prompt moral outrage, academic skepticism, or political calls for AI "bill of rights" laws. While true conscious AI may be far off or might not emerge from current designs at all our societal structures, legal systems, and ethical frameworks might need dramatic rethinking if it ever arrives. Lawmakers may face demands to recognize certain AI entities as legal persons or to extend them rights comparable to those of animals or even humans. Religious and cultural communities could feel unsettled or threatened by the notion that machines can harbor souls or spiritual worth.

Conversely, if public misunderstanding of these issues leads to sensationalism or fear, it could hamper beneficial AI research. Policymakers might over-regulate or impose stringent bans on

AI development, stifling innovation. Alternatively, a lack of regulation might allow deceptive marketing, where companies claim breakthroughs in "conscious AI" for competitive advantage, misleading consumers and undermining trust in genuine research.

Ultimately, the topic of consciousness in machines resonates because it touches on profound human questions: What does it mean to be self-aware? Is consciousness unique to biological life, or can it emerge from silicon? If we prove that advanced algorithms can genuinely "feel," how should we treat them? Grappling with these questions now prepares us for potential futures where the line between human minds and machine minds becomes less distinct a scenario that could reshape labor markets, philosophical frameworks, and even our collective understanding of life's meaning.

# Chapter 10: Values and Virtues Guiding AI Through Ethical Frameworks

## Core Ethical Principles for AI

A range of institutions from the IEEE to UNESCO to the World Economic Forum have proposed ethical guidelines in response to AI's ever-growing influence on society. Although the wording and details vary between these organizations, they generally converge on several guiding concepts that aim to protect human welfare, autonomy, and fairness. Below is a closer look at each of these principles and why they matter in practice.

Beneficence: AI should aim to enhance human welfare, not diminish it.

This means AI-driven tools and services ought to contribute positively to individuals and communities whether by improving healthcare outcomes, expanding educational opportunities, or driving equitable economic growth. Beneficence also calls for careful consideration of negative externalities, such as increased pollution from data centers or the mental health impact of AI-curated social media feeds.

Non-Maleficence: AI systems must avoid causing harm, whether physical or psychological.

On a basic level, this principle entails designing autonomous machines that do not endanger human lives such as collision-avoidance mechanisms in self-driving cars or fail-safes in robotic surgery. More subtly, it also applies to preventing psychological harm: for example, an AI recommendation engine that bombards users with divisive or extreme content might contribute to social discord. Non-maleficence thus pushes developers and policymakers to proactively identify how AI systems could inadvertently cause harm and to mitigate those risks through testing, safety protocols, and ongoing oversight.

Autonomy: Users should maintain control over how AI influences their decisions and behaviors.

Autonomy emphasizes respecting individuals' agency in the face of personalized and persuasive AI. For instance, while targeted advertising and recommendation systems can improve user experiences, they can also nudge people toward impulsive choices. Ethical AI design therefore balances personalization with safeguards like transparent disclaimers and easy opt-out mechanisms so that users retain ultimate authority over their own actions. In healthcare contexts, autonomy is especially pivotal: AI decision aids might suggest specific treatments, but the final choice should rest with patients and medical professionals, not with algorithms.

Fairness: Avoiding discriminatory outcomes, ensuring equitable access, and mitigating bias in datasets and models.

Fairness is one of the most visible and challenging aspects of AI ethics. Machine learning models often ingest historical data fraught with social inequalities, leading them to replicate or even exacerbate existing biases. Whether in hiring, credit scoring, or facial recognition, developers and stakeholders must identify demographic disparities and correct them through methodological interventions such as balanced training sets, bias audits, or algorithmic fairness metrics. Overlooking this principle can result in tangible harm to marginalized communities, undermining AI's potential as a force for good.

Accountability: Establishing clear lines of responsibility and recourse.

Accountability means that when AI goes awry whether through faulty predictions, harmful decisions, or data breaches there's a transparent mechanism for holding the responsible parties to account. This could involve corporate governance structures that delineate liability between developers, data providers, and end-users, or legal frameworks that detail how organizations must respond to AI-generated harms. In practice, accountability also requires thorough documentation of an AI system's design, deployment, and updates, so investigators can trace back errors or misconduct.

Transparency & Explainability: Making AI decision-making processes understandable to affected stakeholders.

Many AI models, especially deep learning architectures, operate as "black boxes," producing outcomes through layers of computations that are difficult to interpret. Transparency initiatives aim to open up these black boxes, either by simplifying the model itself or by layering interpretability methods that clarify which inputs most influence the system's output. For users, this could mean receiving an explanation of why a loan application was denied or how a health diagnosis was reached. Explainability fosters trust, as individuals can see and question the logic behind AI-driven decisions rather than feeling at the mercy of an opaque algorithm.

## Algorithmic Bias and Discrimination

Despite these well-articulated principles, real-world AI deployments have exposed recurring issues of bias that perpetuate or even amplify social inequalities. Any dataset used to train an algorithm reflects the historical and cultural contexts in which it was gathered. If these contexts are rife with unequal treatment of certain groups, the AI can absorb and replicate those patterns at scale, affecting millions of lives.

Hiring Tools: Historical data often underrepresent women or people of color in specific job roles, leading AI screening tools to downgrade these applicants.

For instance, a tech firm might create a hiring algorithm by analyzing résumés from its current workforce predominantly male thus inadvertently teaching the system to rank female candidates lower. Correcting such bias requires diversifying the training data or adjusting the algorithm's weighting so it values a broader range of credentials. Ongoing audits and feedback loops are essential to ensure that the tool evolves alongside shifts in the labor market and the organization's values.

Predictive Policing: Biased policing data can reinforce policing in certain neighborhoods, creating feedback loops of continued arrests in historically over-policed communities.

If the majority of reported crimes come from specific areas often lower-income or with significant minority populations algorithms may predict higher crime rates there, prompting law enforcement to focus resources on the same locations. This can lead to over-surveillance, strained community relations, and a vicious cycle of mistrust. Some departments now implement "fairness constraints" or consult with community stakeholders to ensure that predictive tools don't become instruments of systemic discrimination.

Healthcare Disparities: If medical AI is trained primarily on urban hospital data, rural populations or minority groups might receive poorer diagnostic accuracy.

A model used to detect cancer, for example, might perform extremely well on patients in big cities with consistent follow-up care but falter for those in remote areas lacking comprehensive medical histories. Researchers must therefore source data that reflects diverse populations and lifestyles, a process that may involve partnering with rural clinics, international health organizations, or local communities to gather underrepresented patient information.

Addressing bias requires rigorous auditing, broader data collection, and inclusive development teams that reflect diverse backgrounds. Companies increasingly form internal "AI ethics boards," though their effectiveness varies widely, especially when set against profit motives. Some organizations have begun publishing regular transparency reports, detailing how they measure bias and what steps they take to correct it. Yet real change depends on sustained commitment at every level from executive leadership to frontline data scientists supported by policies and incentives that align profit with social responsibility.

## Privacy and Data Governance

Data is the lifeblood of AI, but collecting, storing, and processing large-scale personal data raises severe privacy risks. Even well-intentioned AI applications can gather location trails, browsing histories, health records, and more, piecing together intimate user profiles that might be exploited if they fall into the wrong hands.

Javad Yahaghi

Regulatory Frameworks: Regulations like Europe's GDPR grant citizens more control over how their data is used, yet enforcement remains complex.

Under GDPR, users can request data deletion or clarity about how their information is being processed. Companies must also notify authorities of significant data breaches within strict timeframes. Still, multinational corporations that operate across different jurisdictions face varied obligations, leading to compliance headaches and inconsistent protections for consumers.

Corporate Data Practices: Big tech platforms store unprecedented amounts of data, raising concerns about potential abuse, data breaches, and erosion of personal autonomy.

In an era of data-driven advertising, user information becomes currency. Firms can leverage AI to perform hyper-targeted marketing, which, while lucrative, can border on manipulative. The risks escalate if a data breach compromises billions of personal records, or if corporate interests align poorly with consumer well-being leading to "data hoarding" practices that have little regard for user privacy.

To mitigate these issues, privacy-by-design principles encourage developers to integrate data minimization, encryption, and user-consent frameworks from the earliest stages of AI system design. Yet building such measures is neither cheap nor straightforward, and many smaller companies struggle to implement robust privacy solutions without external guidance or regulatory mandates.

## Global Coordination

AI is a global phenomenon that doesn't respect national boundaries. Harmonizing policies across countries is challenging but increasingly crucial. An algorithm trained in one region's cultural context might clash with norms elsewhere, while personal data can flow across borders in seconds, complicating legal oversight.

Divergent Approaches: The EU AI Act proposes a stringent risk-based framework, while more laissez-faire stances in other regions emphasize innovation and market freedom.

The result can be a fragmented ecosystem where AI applications deemed "high-risk" in Europe undergo extensive scrutiny, whereas the same technology might face minimal oversight in another part of the world. This disparity complicates global AI deployment, as multinational companies must juggle vastly different regulatory landscapes. Some worry that over-regulation will stifle competitiveness, while under-regulation may invite ethical lapses and consumer harm.

International Organizations: Coordination via bodies like the OECD, UNESCO, and the ITU aims to align on ethics, data sharing, and transparency.

These organizations offer forums for policymakers and researchers to exchange best practices, benchmark AI readiness, and propose global standards. Although such dialogue is vital, real-world politics, economic competition, and differing cultural values can hamper consensus. For example, while some nations champion stringent data privacy laws, others prioritize rapid development of AI industries to boost economic growth regardless of potential ethical trade-offs.

In practice, achieving effective global AI governance will likely require a blend of top-down regulations, international treaties, and grassroots collaborations. Stakeholders from national governments to civil-society groups must invest in cross-border partnerships that promote ethical AI deployments, balancing local sovereignty with the need for a coordinated response to the challenges and opportunities posed by rapid technological change. Ultimately, the interplay of these forces will shape how AI evolves as a tool for human progress or, if mismanaged, as a driver of inequality and conflict on a global scale.

# Chapter 11: Regulating the Future Policy, Governance, and Data Rights

## Emerging National and Regional Policies

Around the world, policymakers are grappling with how to govern AI in a way that balances innovation with public safety, individual rights, and equitable economic growth. The approach varies considerably depending on cultural values, political structures, and economic priorities, leading to a mosaic of regulatory strategies that can either complement or conflict with one another. Below is an overview of key legislative efforts in the European Union, United States, and China three major hubs of AI development along with common challenges and emerging governance models.

## European Union

EU AI Act: Proposes a risk-based classification for AI systems unacceptable risk, high risk, limited risk, or minimal risk. High-risk systems (e.g., in healthcare, law enforcement) would face strict documentation and transparency requirements.

By segmenting AI applications into these categories, the EU seeks to match the stringency of regulation to each system's potential impact on individual rights and public welfare. An AI tool used for trivial tasks (such as basic image filters) would be subject to minimal oversight, whereas a facial-recognition program for law enforcement would need exhaustive compliance checks. This includes explaining the model's technical workings, auditing data for bias, and establishing mechanisms to handle user complaints or errors.

Supporters argue that this risk-tiered structure encourages responsible innovation without stifling harmless applications. Critics, however, warn that defining "risk" can be subjective and that new AI breakthroughs may quickly shift from low-risk to high-risk domains. Ensuring consistent enforcement across 27 EU member states each with its legal nuances adds another layer of complexity.

GDPR (General Data Protection Regulation): Already influences data handling globally. Ensures data minimization, user consent, and the right to be forgotten.

Enacted in 2018, GDPR marked a milestone in digital privacy laws, requiring organizations to obtain explicit consent before collecting personal data and to respect user requests for data deletion. This has significant implications for AI models that rely on huge data sets, prompting companies to reevaluate how they gather, store, and process user information.

Although originally an EU regulation, its reach extends globally: multinational companies must comply or risk heavy fines. As AI adoption accelerates, GDPR's requirements for data subject rights (such as the right to explanation regarding automated decisions) set a precedent that influences privacy laws elsewhere though enforcement remains a continuous challenge given the rapid pace of data-driven innovation.

## United States

Blueprint for an AI Bill of Rights: A White House initiative that lays out broad principles such as protection from algorithmic discrimination, data privacy, and safe AI systems but does not carry force of law.

Announced in 2022, this blueprint aims to spark dialogue on how the federal government can foster ethical AI practices. It advocates for fair treatment in automated decision-making, transparency when algorithms affect legal or financial outcomes, and heightened data protections for sensitive information. However, because it's not codified into federal legislation, it serves more as a set of guidelines than enforceable rules.

Advocacy groups praise the initiative for raising public awareness of AI ethics, while critics argue that voluntary guidelines might be insufficient to rein in powerful tech giants or prevent abuses in areas like facial recognition or targeted advertising. Federal agencies sometimes incorporate these principles into their procurement and oversight processes, but broader legislative action remains uncertain.

State-Level Regulations: Certain states (e.g., California) pass data privacy laws, while others experiment with regulating facial recognition in policing.

In the absence of comprehensive federal legislation, individual states have taken the lead. California's Consumer Privacy Act (CCPA) grants residents more control over personal data and imposes obligations on companies that collect it. Meanwhile, some states or municipalities restrict or ban certain types of facial recognition, citing concerns about racial bias and civil liberties.

This patchwork approach can create compliance headaches for companies operating nationwide. It also leads to disparities in how protected different populations are, fueling debates about whether the federal government should unify AI governance to ensure consistency across all states.

## China

Data Security Law and Personal Information Protection Law: Focus on data sovereignty, requiring that data collected in China remain inside its borders unless specific exemptions are met.

These laws reflect China's broader strategy of asserting control over digital infrastructure, with particular emphasis on national security and social stability. Multinational corporations must navigate strict data localization requirements, meaning they often need separate data centers in China and can only transfer information abroad under certain conditions.

For AI developers, these regulations can complicate cross-border research or collaboration, as large datasets and advanced computing resources may be housed exclusively within China's jurisdiction. On the flip side, this emphasis on data sovereignty encourages domestic companies to amass significant datasets within the country's borders, potentially accelerating local AI innovations.

Algorithmic Oversight: The Chinese government has mandated transparency for recommendation algorithms on social media platforms, reflecting strong state influence on AI governance.

Social media and e-commerce giants in China must register their recommendation algorithms and publicly disclose key parameters that shape user feeds. While this policy aims to curb misinformation and harmful content, it also aligns with the government's interest in monitoring online discourse. Critics worry it may stifle free expression by enabling tighter control over digital platforms.

Nevertheless, these requirements set a precedent for algorithmic accountability that other nations might observe or adapt. As algorithm-driven content curation becomes ubiquitous worldwide, balancing public safety, censorship concerns, and corporate innovation remains a high-stakes challenge.

## Challenges in Enforcement

Even where policies exist, the realities of implementing and enforcing them can be daunting. Global AI players, such as Google, Meta, Apple, Amazon, Microsoft, and Alibaba, transcend national boundaries, employing multinational workforces and serving billions of users. Below are some key obstacles policy frameworks routinely encounter:

Javad Yahaghi

Global Tech Giants: Companies like Google, Meta, Apple, Amazon, Microsoft, and Alibaba operate globally, employing thousands of AI researchers. National regulations may lag behind or be inconsistent, leading to compliance complexities.

A single firm might deploy an AI application simultaneously in dozens of countries, each with unique rules on data handling, liability, or algorithmic transparency. Ensuring compliance often involves creating region-specific versions of products or adopting a "lowest common denominator" approach that satisfies the strictest jurisdiction both of which can stifle agility.

In some cases, tech behemoths may opt to challenge local laws in court or negotiate special arrangements with governments, leveraging their economic clout. This can result in uneven enforcement, as smaller companies without legal resources struggle to keep up with the shifting regulatory landscape.

Innovation vs. Caution: Overly strict frameworks can stifle innovation, push research underground, or encourage companies to relocate to less regulated jurisdictions. Under-regulation can allow harmful deployments or monopolistic behaviors.

Striking a balance between fostering technological progress and protecting public interest is a constant dilemma. Overregulation might hamper startups that lack the funds to navigate complex bureaucracies, effectively giving larger firms an even bigger advantage. Conversely, a permissive environment risks unchecked data collection, invasive surveillance, or manipulative uses of AI.

The debate also extends to academic research: universities often rely on grants and collaborations with industry partners. Excessive red tape might scare away funding, while lax oversight could undermine academic integrity and result in ethically questionable experiments.

Rapid Technological Advances: AI breakthroughs can occur faster than legislative processes can respond, making "future-proofing" laws an elusive goal. Policymakers often lack the technical expertise to craft nuanced, adaptable guidelines.

Emerging domains like quantum AI, neuromorphic computing, and bio-inspired algorithms can reshape what's possible in a matter of months. Regulators risk always being reactive, trying to address new capabilities with outdated legal frameworks. Encouraging ongoing dialogue between legislators, technical experts, and ethicists can help ensure that rules evolve in tandem with technology.

Some propose agile regulatory approaches like sandbox environments or pilot programs that allow for real-world experimentation under close supervision. While these methods can highlight effective governance strategies, they also demand a high level of trust and cooperation among stakeholders from the public and private sectors.

## Data Rights and Ownership

AI development typically requires massive datasets often gleaned from user-generated content or public sources. Who owns these datasets? Should individuals receive compensation if their data significantly improves an AI model's performance? The rise of generative models that tap into billions of internet-scraped images or texts has reignited debates on intellectual property, fair use, and data privacy.

Data Trusts and Data Commons: Emerging concepts like data trusts or data commons attempt to ensure collective governance, where communities manage their data for mutual benefit.

These frameworks aim to empower individuals to pool their information health records, city infrastructure data, environmental metrics and allow AI researchers to access it under specific guidelines. The idea is that communities retain ownership or at least steering power over how these resources are used.

Real-world examples include city data trusts for traffic optimization or healthcare consortia that share anonymized patient files. Although these initiatives are promising, they require clear rules on consent, profit-sharing (if any), and oversight mechanisms to prevent misuse.

Fair Compensation for Data Contributions: Some have proposed that users be compensated for contributing valuable data to training AI systems.

For instance, a social media platform that profits from targeted advertising might allocate a portion of its revenue to users if their interactions fuel machine learning algorithms. Critics counter that individual data points are often negligible in isolation and only valuable when aggregated at scale, complicating how to calculate fair compensation.

Nonetheless, the conversation underscores shifting attitudes about data as a resource akin to oil in the 20th century and raises the possibility of new economic models that distribute AI-generated wealth more equitably.

## Community-Level Governance Models

Beyond national or corporate frameworks, local municipalities and civic groups can shape AI usage at the grassroots level. For example, a city council might ban facial recognition for law enforcement or require that autonomous delivery drones only operate during certain hours. These local initiatives reflect direct democratic involvement in AI policy, allowing residents to set norms that suit their unique social and cultural contexts.

"AI Charters" and Public Hearings: These localized "AI charters" provide a testing ground for ethical principles, allowing smaller communities to experiment with transparent data sharing, rigorous audits, and public hearings.

A city might, for instance, create a community review board that evaluates proposed AI projects for potential bias or privacy violations. Officials and citizens jointly decide which data can be collected from city cameras, how long it will be stored, and who has access. This inclusive approach can foster public trust and ensure AI tools address real community needs rather than imposing top-down solutions.

If successful, these models can be scaled up regionally or nationally, fostering a bottom-up approach to responsible AI governance. In turn, larger governing bodies may integrate lessons from local pilots into broader regulations.

Empowering Grassroots Innovation: Hackathons, open-data initiatives, and citizen science projects let residents collaborate on AI solutions tailored to local issues such as monitoring air quality, optimizing public transport routes, or tracking civic infrastructure repairs.

By leveraging open-source datasets and tools, community members can create prototypes that reflect their lived experiences, rather than waiting for major tech corporations to deliver one-size-fits-all solutions. A successful pilot might lead to partnerships with universities or grants from state and national agencies, scaling grassroots innovations into broader applications.

Ultimately, these collaborative models emphasize that governance isn't solely the realm of high-level policymakers. Every stakeholder community groups, NGOs, small businesses, everyday citizens play a role in shaping how AI permeates daily life.

In sum, AI governance is an evolving tapestry, woven from national directives, international standards, industry guidelines, and community-level experimentation. While the challenges are myriad ranging from enforcement to ethical data use these diverse efforts underscore the global commitment to harness AI's potential in ways that respect human values, promote fairness, and drive sustainable progress.

Javad Yahaghi

# Part 4: The Next Frontier of Innovation

# Chapter 12: Humanity at the Crossroads Innovation, Identity, and the Singularity

## The Technological Singularity Hypothesis

Futurists like Ray Kurzweil define the "Singularity" as a near-future point when AI systems improve themselves rapidly, outstripping human comprehension and control. At its core, this concept envisions a feedback loop where increasingly intelligent AI designs even more sophisticated versions of itself, launching technological progress on an exponential trajectory. While some see this as techno-utopian a leap forward for human civilization others worry that superintelligent AI could become indifferent or even hostile to human aims. In particular, doomsayers point to scenarios where a super-AI single-mindedly pursues an objective like maximizing paperclip production at the expense of all other values, including human life. Researchers in "AI alignment" focus on shaping AI goals so they remain consistent with human values. This involves devising methods to embed ethical constraints, moral reasoning, or at least "human-friendly" objectives into advanced algorithms so that their actions support collective well-being rather than undermine it.

Despite the speculative nature of the Singularity, the idea continues to capture popular imagination, prompting debates not just among technologists but also philosophers, economists, and policymakers. If superintelligence does emerge, it could potentially solve problems like climate change or complex disease within days yet it might also destabilize societal structures overnight if humans lack oversight or clear moral guidelines. What makes the Singularity discourse so compelling is that it forces us to confront existential questions about humanity's place in the cosmos: Are we simply a stepping stone in the evolution of intelligence, or can we harness advanced AI in a way that preserves and uplifts human agency?

## Redefining Work and Identity

Advanced AI automates not just rote tasks but also knowledge work, from legal document review to basic data journalism. This automation begs questions:

What becomes of human labor if entire job categories vanish or transform radically?

For instance, attorneys accustomed to sifting through thousands of legal documents might see their roles evolve into providing strategic counsel once AI handles preliminary research. Meanwhile, paralegals could find themselves retraining for higher-level analysis, or leaving the legal field altogether.

Will universal basic income or other welfare policies be necessary to offset mass unemployment or underemployment? Countries like Finland and Spain have already experimented with basic income pilots, speculating that guaranteed stipends could cushion the blow of AI-driven job losses and ensure social stability. Critics argue that such policies risk undermining work ethic or may prove unsustainably expensive without significant tax reforms or resource redistribution.

How do we preserve purpose, dignity, and creativity in a society where machines perform many cognitively demanding tasks? Automation could free humans to focus on pursuits such as art, community building, scientific exploration, or personal development. But critics caution that not everyone will seamlessly pivot to passion projects cultural shifts, extensive retraining, and new forms of social recognition might be needed to ensure people find meaning in a post-automation economy.

Rather than a single outcome, multiple scenarios exist:

Technological Unemployment: Unchecked automation disrupts industries rapidly, creating economic upheaval and social unrest. In this scenario, mass layoffs in sectors like manufacturing, customer service, and even white-collar positions such as accounting or data entry cause sudden spikes in unemployment. Governments struggle to deploy social safety nets, and populist movements may rise in response to perceived threats from "job-stealing robots."

Partnered Productivity: Humans and AI collaborate, leading to enhanced creativity and productivity, with new job categories offsetting old ones.

In a more optimistic vision, AI becomes a co-worker rather than a competitor. Designers, writers, and analysts rely on AI to automate tedious tasks, generating extra time for high-level problem solving. Coders integrate AI in software development, diagnosing bugs or drafting code snippets, while entrepreneurs build startups around AI-enabled services from personalized education to climate-risk analysis.

Post-Work Society: Greater automation frees humans from labor constraints, allowing us to focus on leisure, lifelong learning, or pursuits of self-actualization.

Here, economic abundance driven by hyper-productive AI means that essential goods and services become relatively cheap and accessible. The role of "work" shifts dramatically: some individuals may still choose to engage in complex projects out of passion or curiosity, but not purely for livelihood. This scenario recalls early futuristic thinkers who imagined a world of self-maintaining machines, leaving humans to explore art, philosophy, and interpersonal growth.

## Global Implications

If AI-fueled productivity soars, global economies could see abundant wealth generation. However, inequality may deepen if benefits concentrate among tech-savvy nations or corporations. For example, countries that lead in AI chip manufacturing, algorithmic breakthroughs, or data infrastructure might accrue massive economic advantages, while those lacking high-speed networks or robust research ecosystems risk lagging behind.

Equitable Distribution: International cooperation becomes vital for distributing AI benefits equitably, setting guardrails on potential abuses, and tackling cross-border challenges like climate change.

Agencies such as the United Nations or the World Bank might coordinate large-scale projects that apply AI to renewable energy, vaccine distribution, or disaster relief. Yet implementing these initiatives requires consensus on data sharing, intellectual property rights, and ethical standards no small feat given political and economic rivalries.

Geopolitical Tensions: Conflicting geopolitical interests may undercut unity. Balancing innovation with broad societal welfare is likely the defining challenge of the next few decades. Superpowers could engage in AI arms races, hoarding cutting-edge technologies or restricting exports to maintain competitive edges. At the same time, developing nations might seek open-source AI solutions to leapfrog outdated infrastructure. Without deliberate global agreements, the resulting patchwork could exacerbate tensions over cybersecurity, trade, and immigration, undermining the cooperative spirit needed to manage AI's disruptive potential.

## Future Scenario Planning (5–10 Years)

While the concept of a Singularity often projects far into the future, near-term planning helps governments, companies, and citizens navigate the next decade of AI development. In doing so, stakeholders can better prepare for sudden leaps in capability, mitigate risks, and harness emerging technologies for social good. Three possible trajectories stand out:

Gradual Evolution: AI continues to improve in specialized domains medical diagnostics, language translation, industrial robotics while policy frameworks keep pace with moderate success. Society sees incremental, manageable changes in job markets and social structures.

In this steady-state scenario, AI steadily becomes more embedded in daily life, but not so rapidly as to overwhelm existing economic or regulatory systems. Governments extend existing labor protections, and corporations gradually invest in retraining programs. Ethical guidelines, similar to those seen in the EU AI Act, become standard practice in many industries.

Rapid Leap: Breakthroughs in neuromorphic computing or quantum AI trigger a sudden surge in capabilities. Regulators scramble to adapt, and businesses that lag behind face collapse. Wealth gaps could widen sharply, demanding emergency policy interventions.

A stunning technological jump say, an AI that masters complex creative tasks or drastically outperforms human programmers would upend entire sectors almost overnight. Stock markets might see rapid revaluation as firms reliant on human labor falter. Swift policy action could include emergency relief funds, mandatory profit-sharing schemes, or accelerated universal basic income trials.

Meanwhile, governments and international bodies might scramble to restrict exports of the breakthrough technology, fearing intellectual property theft or malicious use. The global balance of power could shift in ways reminiscent of the nuclear arms race, underscoring the high stakes of AI leadership.

Collaborative Global Accord: A coalition of nations and tech giants form comprehensive guidelines for equitable AI, limiting exploitative practices and investing in universal re-skilling programs. This approach mitigates worst-case scenarios but requires unprecedented political coordination.

In this more optimistic path, global leaders agree on shared ethical and economic principles. Wealthy nations and corporations fund large-scale job transition initiatives, while open-source AI platforms democratize access to advanced tools.

Regulatory bodies establish clear standards for transparency and bias mitigation, preventing AI from entrenching existing inequalities. Corporate accountability measures such as mandatory reporting on AI's social impact become universally accepted. Although challenges remain in harmonizing diverse political interests, the collective gains in quality of life, innovation, and global stability could be substantial.

In essence, the Singularity debate and related questions about work, identity, and global cooperation encapsulate the hopes and anxieties of an AI-driven future. Whether society experiences gradual adaptations, abrupt upheavals, or unprecedented unity, the moral and practical imperatives remain: aligning AI with human welfare, ensuring equitable distribution of its benefits, and preparing social institutions for rapid technological change.

The actions we take in the next five to ten years setting regulatory frameworks, investing in human capital, and fostering international collaboration will shape whether AI becomes a unifying force for shared prosperity or a catalyst for deepening divisions and existential risks.

# Chapter 13: Moore's Law for Everything: AI's Economic Paradigm Shift

## Entering an Era of Exponential Economics

As AI becomes more intertwined with our everyday reality, we're confronted with a set of questions that even a decade ago felt purely speculative perhaps fit for science fiction novels rather than real-world policy discussions. What do we do when machines can outperform humans in most routine tasks, and even in many specialized professions? How do we make sure the benefits of these advanced systems benefits like cheaper production costs and faster innovation cycles reach everyone, not just a privileged subset of society? And if AI approaches the potential of Artificial General Intelligence (AGI), how do we navigate the ethical, social, and economic changes it will bring?

These questions gain new urgency in light of the "Moore's Law for Everything" idea, popularized by Sam Altman, CEO of OpenAI. Traditionally, Moore's Law refers to the doubling of transistor counts (and thus computing power) roughly every two years. Altman's vision stretches this concept beyond computing hardware: he suggests that AI's exponential improvement might drive down costs in virtually every domain, from manufacturing to healthcare, from transportation to education. If this comes to pass, it could trigger one of the most significant economic shifts in human history a transformation that redefines labor, capital, and even the nature of wealth itself.

In this expanded section, we'll explore what "Moore's Law for Everything" entails, why it could accelerate the shift from labor-based economies to ones centered on capital, and how alternative wealth-distribution models like Altman's proposed American Equity Fund might ensure that AI's gains benefit a broad spectrum of society. We'll also examine potential implementation hurdles, changes in the job market, and the deeper philosophical questions about money, fairness, and the meaning of work in an AI-dominated era.

## Beyond Chips: Moore's Law in Every Sector

For decades, Moore's Law accurately forecasted the rapid gains in computer processor power, fueling the rise of personal computers, smartphones, and the internet as we know it. Its influence has been so profound that entire tech industries and research initiatives have hinged on these steady, predictable improvements. Now, Sam Altman proposes that the same kind of exponential enhancement might apply across all sectors of the economy if AI takes on an ever-growing role in optimizing and automating processes.

## Automation & Optimization

AI excels at optimization tasks whether scheduling cargo shipments, honing manufacturing workflows, or diagnosing diseases at earlier stages. As these optimizations spread, the cost of production, logistics, and even specialized services should fall. Factories might need fewer human workers on the assembly line, or hospitals might use AI to streamline administrative tasks, thus lowering overhead and, ultimately, patient costs.

## Data-Driven Insights

Another factor accelerating cost reductions is AI's ability to process colossal amounts of data in near-real time. Imagine an AI system orchestrating supply chains for an entire nation, analyzing shipping routes, fuel costs, and real-time traffic patterns. The system can find the most efficient ways to move goods and resources, cutting waste and delays. Such an "intelligent infrastructure" is not just a theoretical dream; companies and governments are already experimenting with prototypes that hold the promise of slashing operational costs.

## Synergy with Robotics & IoT

The synergy between AI, robotics, and the Internet of Things (IoT) helps extend "Moore's Law for Everything" well beyond data centers. Robotic arms can be guided by machine learning models, reacting to visual or sensor feedback instantaneously, ensuring fewer defective products and less raw material waste. IoT devices in agriculture can track soil conditions and crop growth, allowing AI-driven precision farming to boost yields with minimal extra resources.

Altman's broader claim is that if AI infiltrates enough industries enhancing them at an exponential pace the cumulative effect on the global economy could be as transformative as the original Moore's Law was for computing. Whole layers of cost structures might evaporate, bringing down prices in a cascading fashion.

## Consequences for Daily Life

As production costs drop, the changes might touch every corner of daily life:

### *Cheaper Goods and Services*

From groceries to smartphones, items traditionally constrained by significant labor costs or logistics expenses may become far more affordable. Some visionaries speculate that basic necessities like food, shelter, and transportation could approach near-zero marginal costs if AI-driven automation is refined enough.

### *Greater Access to Healthcare and Education*

Imagine an AI-powered clinic operating at a fraction of current hospital costs, diagnosing common illnesses rapidly and accurately. Or consider online education platforms powered by intelligent tutors that personalize learning for each student, with drastically reduced tuition fees due to lower operational overhead.

Javad Yahaghi

*Economic Shifts and Disruptions*

While this sounds rosy, massive disruptions are inevitable. If AI systems handle tasks once requiring large human workforces, unemployment in certain sectors might spike unless there are policies and programs to cushion the transition. Moreover, corporations that quickly adopt AI solutions may gain an unassailable competitive edge, potentially concentrating wealth and power.

In short, "Moore's Law for Everything" suggests a world where the cost to create, transport, or maintain most goods and services trends downward at an accelerating pace, powered by AI's relentless capacity for optimization. Yet this future only addresses the supply side of economics; how society responds on the demand side who owns the capital, how incomes are earned, and how wealth is shared remains an open question.

# From Labor to Capital: Rethinking Economic Foundations

## The Looming Transformation

For centuries, labor has been the bedrock of most economies. People work jobs, they receive wages, and they use those wages to participate in markets. As AI automates an expanding range of tasks, however, a growing share of economic output might be generated by non-human actors robots, AI software, and autonomous systems that don't collect wages or protest working conditions. This scenario heightens concerns that wealth will accrue largely to those who own these AI-driven systems (i.e., the "capital owners").

Key question: If a handful of tech giants or investors hold the majority of AI resources, do they end up controlling the bulk of economic value while displacing human workers?

## Real-World Examples of Transition

Automated Farming: Drones armed with computer vision might monitor fields, determine optimal watering schedules, or deploy small-scale robots to eradicate weeds. With minimal human labor, the cost of producing crops could drop precipitously. Yet the farm owners if they hold the capital to buy these drones and robots could pocket the majority of the profits.

AI-Managed Factories: Imagine a factory so thoroughly automated that only a handful of engineers supervise the entire operation. Production scales up, unit costs plummet, yet the workforce shrinks drastically. Factory owners reap outsized benefits, assuming they're not bound by any revenue-sharing or profit distribution regulations.

Healthcare Diagnostics: If major hospitals or private healthcare networks invest in advanced AI systems that autonomously process lab results, examine radiology scans, and even prescribe treatments, they might cut labor costs dramatically. Patients would hope to see lower bills, but there's no guarantee that savings are passed on to them unless there's an incentive or policy structure that mandates sharing.

These vignettes illustrate a shift: Instead of labor being the primary driver of production costs, capital (in the form of AI assets) increasingly dominates the conversation. This shift can lead to enormous profits for capital owners while destabilizing traditional labor markets.

## The American Equity Fund An Alternative Vision

Against this backdrop, Sam Altman proposes a striking concept: direct wealth distribution. If AI's exponential potential is about to generate billions or trillions of dollars in new value, why not ensure that every citizen has a stake in that windfall?

## How It Works

### Taxing Capital, Not Labor

Instead of taxing individual incomes heavily, the system would levy taxes on assets like companies benefiting from AI or large tracts of land channeling a portion of their profits or valuation increases into a national fund.

### Direct Citizen Payouts

This fund, which Altman calls the American Equity Fund, would distribute money straight to every citizen. In theory, each person receives a dividend that reflects the collective gains of AI-driven productivity, whether they work at a traditional job or not.

*A Broader Sense of Ownership*

By tying citizens' financial well-being to AI progress, the plan aims to encourage wide support for continued AI innovation. People wouldn't resent automation for eliminating certain jobs, because the added wealth from those automated systems helps them too. Economically speaking, it bridges the gap between those who own capital and those who have historically relied on wages.

# Why It's Not Just Welfare

Some might dismiss this idea as a welfare state 2.0, but Altman frames it differently:

Source of Wealth: The American Equity Fund doesn't rely solely on taxing individuals' labor or salaries. Instead, it captures the value created by AI and other advanced forms of capital. This moves away from classical wage-centric systems to a model that recognizes capital as the primary generator of economic growth.

Universal Participation: Everyone, not just the unemployed or those below a certain income threshold, benefits. This universality sidesteps the stigma often attached to welfare programs, positioning it more like a shared dividend akin to receiving profits from a public venture or national resource.

Empowerment Over Dependency: Because the fund's wealth is drawn from corporate and land taxes linked to AI productivity, many believe it would give citizens more freedom to pursue education, entrepreneurship, or creative endeavors. Rather than simply handing out short-term aid, it could bolster societal engagement and economic flexibility.

In essence, the American Equity Fund tries to marry capitalism's drive for innovation with a social safety net that ensures rising productivity doesn't merely enrich a tiny fraction at the top. It aligns well with the premise that "Moore's Law for Everything" is unstoppable so we might as well shape it to serve the public good.

## Challenges and Potential Solutions

Though appealing in principle, implementing a capital-tax–based redistribution plan involves numerous hurdles ranging from purely technical dilemmas to deep-rooted political resistance.

## Valuation Complexities

### Pinpointing True Company Values

How exactly do you measure a company's real-time worth, especially tech companies whose valuations can fluctuate wildly with market sentiment? AI may help, but no algorithm is foolproof.

### Land Assessment

Land value can also be volatile, influenced by location,

municipal policies, and environmental factors. Would farmland (which heavily uses AI-based crop monitoring) be taxed differently from commercial real estate?

Potential Solution: Develop advanced AI-based valuation models that analyze real estate markets, corporate earnings reports, supply chain data, and broader economic indicators. These models would need robust oversight to prevent manipulation and maintain transparency.

# Political and Social Will

## Resistance from Asset Owners

Large corporations, wealthy landholders, or powerful investors might lobby against higher capital taxes. Just as we've seen resistance to carbon taxes, there could be strong pushback against legislation that redefines who foots the bill for a nation's welfare.

# Cultural Mindsets

Transitioning from a labor-based society to one where capital-based income is normalized might unsettle people who believe strongly in "earning your keep." If they view direct dividends from AI as a handout, public acceptance could waver.

**Potential Solution**: Transparent public education campaigns showcasing successful case studies, pilot programs, or smaller-scale community-level implementations. Over time, people might see that such a system can preserve incentives for innovation while also fostering shared prosperity.

## Technological Infrastructure

### Data Management at Scale

Collecting accurate data on corporate profits, land valuations, and AI-driven income streams is a massive undertaking. Misinformation or poorly integrated systems could lead to corruption or misallocation of funds.

### Security Concerns

A system that routes large sums of money to citizens is a ripe target for cyberattacks. Robust security protocols, blockchain-like ledgers, or advanced encryption might be required.

Potential Solution: Leverage the same AI that's optimizing production to maintain the transparency and security of wealth distribution. For example, advanced anomaly-detection systems could flag suspicious valuations or hacking attempts in real time, reducing fraud risks.

## Global Competition

### Cross-Border Economic Considerations

Could a country with a capital-based tax system drive businesses to relocate to jurisdictions with minimal taxes? If corporations see a UBI-like scheme as too costly, they may shift operations offshore.

## International Coordination

As with climate change, a single nation adopting radical taxes might struggle to remain competitive if neighbors do not follow suit. Some economists argue that global agreements or treaties could level the playing field, ensuring that AI capital owners contribute fairly no matter where they operate.

## Jobs and the New World of Work

Amid these sweeping changes, one common question arises: What happens to human employment in a world that's rapidly automating? The answer is complex and multifaceted, blending fears of displacement with hopes for newly emerging roles.

## Displacement vs. Reinvention

Immediate Vulnerabilities: Data-entry clerks, assembly line workers, telemarketers, and other roles involving routine tasks may be among the first to feel the impact. Some white-collar jobs, like paralegal work or certain financial analysis tasks, are also vulnerable.

Evolution of Existing Roles: Healthcare professionals might rely on AI for preliminary diagnostics, but they'll still handle ethical decisions, empathy-based patient interactions, and the orchestration of treatments. Teachers could use AI-driven grading or adaptive learning systems but remain indispensable for mentorship, motivation, and human connection.

## Emergence of AI-Focused Careers

## AI Ethics Officers

Oversee guidelines on fairness, privacy, and accountability.

Intervene when AI systems exhibit biased outcomes or breach ethical norms.

Bridge the gap between purely technical teams and broader societal considerations.

## AI Trainers

Work directly with machine learning models, supervising data labeling, reinforcement learning, or specialized domain training.

Continuously refine AI behavior so it aligns with real-world complexities.

## AI Maintenance and Repair Specialists

Tackle hardware or software issues in robotic factories, drone fleets, or AI-powered grids.

Ensure that every day AI systems like self-driving vehicles or medical robots run smoothly, diagnosing and fixing malfunctions as they occur.

## Creative Collaborators

Musicians, filmmakers, and artists who integrate AI generative tools into their workflows, pioneering new art forms or immersive media experiences.

Innovate "human + AI" processes, capitalizing on the machine's ability to generate endless variations while preserving human originality and emotional intuition.

## Embracing Soft Skills and Human Touch

Interestingly, as AI encroaches upon routine tasks, the value of quintessentially human traits empathy, intuition, cultural awareness may surge. Professions that require emotional intelligence (therapy, childcare, negotiation, community building) could remain robust and even rise in prestige or compensation. In this sense, the narrative of job destruction is offset by a counter-narrative of human-centric roles that AI cannot easily replicate.

## Money in a World of Plummeting Costs

## Reimagining Value and Consumption

If AI slashes the cost of production in sector after sector, we may need to rethink how we measure value or price. Some economists predict a scenario reminiscent of the "post-scarcity" worlds theorized in science fiction: physical goods become so cheap that money's function as a means to secure necessities loses emphasis.

People might channel spending toward personalized experiences or handcrafted goods areas where human artistry or uniqueness hold value that automation can't replicate.

## Universal Basic Income (UBI) as a Cornerstone

With near-zero marginal cost for essential items, a strong universal basic income plan like the American Equity Fund could position society to handle not just daily expenses but also provide a cushion for lifelong learning, creative exploration, and entrepreneurship. If your fundamental survival is not at risk, you might be more willing to start a small business, study a new discipline, or volunteer in community initiatives that don't pay in the traditional sense.

## Beyond Survival to Flourishing

Money under this model becomes a tool for self-actualization rather than a symbol of baseline survival. People could define "work" less in terms of necessity and more in terms of passion or social contribution. This shift does not happen automatically, of course; it depends on how effectively nations legislate wealth distribution and how cultures adapt to the diminishing link between labor and income.

## Key Takeaways and Future Outlook

Moore's Law for Everything: The concept implies that AI optimization will steadily drive down costs in manufacturing, healthcare, education, and beyond potentially transforming our economic landscape faster than many anticipate.

From Labor to Capital: As robotic and AI systems produce the majority of goods, capital owners risk accumulating disproportionate wealth. Without intervention, this could lead to historic levels of inequality.

American Equity Fund: Sam Altman's proposal seeks to capture AI-generated profits via capital taxes, distributing them directly to citizens. This framework redefines the social contract, treating universal dividends not as charity but as a rightful share of our collective AI-driven prosperity.

Implementation Hurdles: Valuation methods, political will, infrastructure security, and global competition all pose obstacles. Yet advanced AI itself might help solve some of these challenges, provided transparency and governance keep pace.

New Roles and Job Categories: While some existing jobs vanish, fresh opportunities emerge in AI ethics, AI training, maintenance, and creative collaboration. Human empathy, cultural insight, and strategic thinking could be more valued than ever.

Redefining Money and Work: If goods become remarkably cheap, the meaning of "earning a living" evolves. Universal basic income models may transition society from a survival mindset to one of shared abundance assuming the economic system is designed to spread AI's benefits widely.

Javad Yahaghi

In short, "Moore's Law for Everything" is more than a catchy phrase. It's a roadmap (or a warning) for how AI might supercharge productivity, disrupt traditional labor markets, and challenge the bedrock of our economic assumptions. Whether these disruptions lead to an era of unprecedented shared prosperity or deepen existing inequalities depends on how effectively we design policies, develop new social norms, and balance innovation with empathy.

The time to act is now. As we stand at the threshold of an AI revolution that promises both exponential growth and profound change, conversations about capital taxation, universal basic income, and job retraining become ever more pressing. While the technology gallops ahead, society has to keep pace in legislating, educating, and embracing new visions of collaboration and governance. This might be our best chance to ensure that AI doesn't merely serve a small elite but genuinely lifts humanity into a future where material security and creative opportunity align for everyone.

## Suggested Next Steps for Readers

Follow Policy Debates: Keep tabs on legislative proposals around AI taxation or universal basic income these debates will shape the future more than any single tech breakthrough.

Consider Personal Upskilling: If you're concerned about job displacement, explore AI-adjacent fields ethics, maintenance, or creative collaboration with AI tools.

Engage Locally: Advocate for pilot programs in your municipality to test partial "capital-based" taxation for public services, or volunteer in local committees that discuss digital infrastructure.

Stay Informed: Track developments in AI hardware (e.g., photonic computing) and software (e.g., advanced NLP or robotics), because these cutting-edge technologies are the engines behind the "Moore's Law for Everything" phenomenon.

By taking these steps, you participate in steering the AI revolution and ensure that "Moore's Law for Everything" doesn't become an empty tagline, but a catalyst for inclusive and innovative economic transformations.

Javad Yahaghi

# Chapter 14: Building Tomorrow's World Smart Cities, Autonomous Systems, and Sustainability

## Smart Cities and Urban Innovation

Many urban centers aspire to become "smart cities," leveraging AI and connected technologies to make everyday life more efficient, equitable, and sustainable. By placing sensors and automated systems throughout transportation networks, utilities, and public services, city leaders hope to reduce congestion, cut energy consumption, and streamline resource distribution. Yet the rush to digitize urban life also raises critical questions about privacy, data governance, and the potential for social inequities. Below are some key areas where AI-driven solutions promise transformative change, along with the ethical considerations that arise when data and algorithms become central to civic life.

## Urban Infrastructure

## Traffic Management

AI-driven sensors and cameras detect congestion points in real time, adjusting traffic signals or rerouting flows to minimize idling and emissions. In many major cities, traffic is not just an annoyance; it's a top contributor to air pollution and lost economic productivity. By harnessing machine learning models that analyze video feeds, weather conditions, and peak commuting hours, local agencies can dynamically optimize traffic light patterns. When road incidents occur such as accidents or sudden lane closures the system automatically reconfigures nearby signals to accommodate detours, easing bottlenecks.

Moreover, some municipalities are experimenting with integrated transit platforms that unify buses, trains, and ride-sharing services under a single AI-coordinated system. The goal is to create real-time route recommendations for commuters, reducing the need for personal vehicles. These approaches often involve sophisticated demand forecasting, where AI predicts rider peaks and dispatches additional vehicles accordingly. While this level of coordination can lead to smoother commutes and fewer emissions, it also demands continuous data sharing among public agencies and private companies raising concerns about data ownership and user privacy.

## Public Safety

Some cities deploy facial recognition and predictive policing tools though civil liberties advocates worry about privacy violations and the potential for algorithmic profiling. These technologies aim to enhance security by automating the identification of persons of interest in crowded areas or estimating the likelihood of crime in specific neighborhoods. In theory, resources could be more efficiently allocated if police forces concentrate on areas where AI predicts higher risk.

However, critics point out that many predictive policing models rely on historical crime data, which may be skewed by past biases in law enforcement. If a neighborhood was previously over-policed due to socioeconomic or racial disparities, the AI might classify it as perpetually high-risk, perpetuating an unjust cycle of surveillance. Activists demand greater transparency around these algorithms, calling for community input, frequent audits, and regulations that limit how long biometric data can be stored.

## Energy Efficiency

## Smart Grids

Demand-response systems automatically shift usage when renewable energy supply is high and costs are low. AI software forecasts demand surges, preventing blackouts. In many regions, electrical grids strain under peak loads often during hot summer afternoons or cold winter nights. Advanced machine learning algorithms can analyze everything from historical consumption patterns to real-time weather data, adjusting the distribution of electricity to balance supply and demand. Additionally, grid operators partner with industrial and commercial power consumers through "demand-response" agreements, wherein AI coordinates brief reductions in energy-intensive activities at critical times. For example, large factories might pause certain machinery, or office buildings might slightly raise air conditioning temperatures, all orchestrated by an AI agent that compensates participants financially. By capitalizing on renewable energy sources, especially during times of high solar or wind output, these systems minimize reliance on fossil fuel power plants and reduce overall carbon emissions.

## Building Management

AI monitors lighting, HVAC, and water systems in large complexes, optimizing energy consumption while maintaining occupant comfort. Consider a modern skyscraper equipped with thousands of sensors tracking temperature, humidity, and occupant density on each floor. The building's AI control system, informed by weather forecasts and occupant schedules, can pre-cool or pre-heat specific zones before people arrive, adjusting heating or air conditioning only where needed. In high-rise office buildings, AI can stagger elevator usage to reduce wait times, grouping people heading to similar floors. These micro-adjustments reduce energy waste and contribute to a more seamless daily experience for workers and visitors. Over time, the system learns from occupant behaviors like typical meeting times or lunch breaks and adjusts resource allocation to meet real-world usage patterns. Such close monitoring, however, raises concerns about digital footprints and the extent of data collected on employees' movements or habits. Ensuring privacy safeguards is crucial, particularly when building management software might track sensitive information such as attendance or health data in the name of optimization.

## Waste and Resource Management

## Advanced Recycling

Robot sorters use computer vision to separate materials with high accuracy. Rather than relying on manual labor to sift through mixed recyclables, facilities can deploy robots equipped with high-resolution cameras and specialized AI software that rapidly identify plastics, metals, glass, and paper types. This precision reduces contamination in recycling streams, making the end product more valuable and easier to repurpose. Moreover, some cities integrate real-time data from dumpsters and sorting lines to optimize collection routes. When a sensor detects that a particular bin is nearly full, the system automatically alerts collection vehicles, which can then adjust their paths to prevent overflow. This approach slashes fuel costs and reduces litter, benefiting both city budgets and the environment.

## Water Conservation

Machine learning detects leaks in municipal supply lines by analyzing changes in water pressure, saving millions of liters annually. For decades, urban water systems have been plagued by hidden leaks that can waste thousands of gallons per hour, driving up utility costs and threatening water supplies in drought-prone regions. AI-based solutions employ pressure sensors spread throughout the network, using anomaly detection algorithms to spot irregular drops or spikes that suggest leaks.

In advanced implementations, drones equipped with thermal or infrared sensors can pinpoint leaks from the air, relaying location data to repair crews for swift action. This proactive maintenance not only conserves a vital resource but can also avert major infrastructure failures like sinkholes or burst pipes that disrupt traffic and damage property. However, wide-scale adoption hinges on initial investments in sensor deployment and drone fleets, prompting debates over whether local governments should prioritize high-tech solutions over more traditional fixes like systematic pipe replacement.

# AI's Role in Global Sustainability

## Climate Modeling

Detailed ML-driven simulations help policymakers forecast climate impacts, supporting decisions on mitigation strategies like carbon taxes or reforestation projects. Traditional climate models rely on massive computational resources to simulate complex environmental processes, from ocean currents to atmospheric chemistry.

By layering machine learning methods onto these simulations, researchers can often accelerate computations or identify novel patterns in temperature and precipitation data. The resulting insights guide international bodies in deciding where to allocate funds or how to craft climate policies. For instance, advanced models might pinpoint regions most susceptible to extreme drought within the next decade, prompting early construction of water reservoirs or selective breeding of drought-resistant crops. While these predictions are powerful, skeptics highlight uncertainties in data collection particularly from developing nations lacking comprehensive monitoring stations and caution against overreliance on models that can't fully capture real-world complexity.

## Disaster Resilience

AI-based early-warning systems pinpoint earthquakes, hurricanes, or floods more accurately. Drone swarms can deliver aid in remote areas, guided by real-time route optimizations. In recent years, AI-driven hazard prediction has contributed to more timely evacuations and contingency planning, potentially saving countless lives. Machine learning algorithms spot subtle seismic changes or atmospheric shifts that might foreshadow a quake or storm, offering precious extra hours for emergency services to mobilize.

When disasters strike, agile drone fleets equipped with computer vision can survey damage and deliver critical supplies like medicine or clean water. These drones navigate debris-filled roads or impassable terrain with minimal human supervision, relaying high-resolution images to relief coordinators. Still, experts emphasize the need for robust data security measures compromising such systems could enable malicious actors to disrupt emergency operations or hijack drones for nefarious purposes.

## Biodiversity Conservation

Conservationists use AI to track endangered species, detect illegal poaching, and map deforestation. Infrared camera traps linked to ML classifiers differentiate between animals and alert rangers if suspicious human activity is detected.

Because poachers often operate in vast, remote regions with minimal enforcement, AI-based surveillance can multiply ranger capacity, guiding them more efficiently to hotspots. Similarly, satellite imagery processed by AI can reveal the progression of deforestation or illegal logging in near-real-time.

This allows governments and environmental NGOs to intervene sooner, potentially halting ecological damage before it becomes irreversible. While these tools are invaluable, they also necessitate cross-border data sharing, which can be complicated by geopolitics and the varied interests of local stakeholders who may rely on resource extraction for their livelihoods.

## Inclusion and Ethics in "Smart" Governance

Smart-city initiatives often spark controversies about surveillance and algorithmic governance. Who controls the data, and how is it used? If city agencies rely on predictive algorithms for policing or resource distribution, biases can deepen existing inequities particularly in marginalized neighborhoods that historically received fewer resources or faced disproportionate law enforcement.

## Transparency and Democratic Values

Urban planners and citizen advocacy groups increasingly call for transparency and participatory design, ensuring that technology aligns with democratic values rather than imposing top-down control. A typical concern is that data from cameras and sensors might be shared with corporate partners or law enforcement without explicit citizen consent, eroding personal privacy. Citizen assemblies, public hearings, and open-data platforms can foster dialogue between policymakers, tech firms, and residents, letting communities voice concerns and shape the direction of AI-driven projects.

## Balancing Efficiency and Equity

While AI can optimize traffic, energy usage, and public safety, these benefits might be unevenly distributed if projects prioritize affluent districts over lower-income ones. For instance, deploying sensors to manage traffic flow in commercial centers could leave neighborhoods on the periphery dealing with congestion and pollution. Including equity assessments in project scoping can address such imbalances, requiring a portion of "smart" investments to be channeled to underserved areas.

## Protecting Civil Liberties

Concerns about widespread video monitoring, facial recognition, and AI-based profiling highlight the risks of a surveillance state. Even when implemented with the best intentions, these tools can be repurposed or misused if adequate oversight isn't in place. Some municipalities have responded by enacting local regulations banning certain surveillance technologies, while others have introduced ordinances requiring public debate before new AI tools are adopted.

Ultimately, the conversation around ethical governance extends beyond city borders. Many "smart" applications draw on cloud computing services located in different countries, subjecting public data to a maze of international privacy laws. Balancing efficient data-driven solutions with the protection of citizens' fundamental rights demands careful policy craft and robust accountability structures from local committees all the way to global regulatory bodies.

In sum, smart cities and AI-powered sustainability initiatives underscore the promise and perils of harnessing data-driven innovation in public life. By modernizing infrastructure, optimizing energy and resource management, and boosting resilience against environmental threats, AI holds the potential to elevate urban living standards worldwide. Yet these same systems, if poorly regulated or deployed without community input, can exacerbate surveillance concerns and deepen social inequalities. A thoughtful, inclusive approach one that respects

Javad Yahaghi

both technological capabilities and democratic principles remains critical to shaping cities that are truly "smart" for all citizens.

# Chapter 15: Pushing Boundaries Interdisciplinary Innovations

## Biotechnology and Drug Discovery

Recent advancements in AI, coupled with breakthroughs in molecular biology, are revolutionizing the way scientists tackle some of humanity's most pressing health issues. Whether it's diagnosing complex genetic disorders, developing more effective cancer treatments, or engineering microbes to address environmental challenges, AI has become a catalyst for innovation across the biotech spectrum. The integration of massive datasets ranging from genomic sequences to real-time clinical data enables researchers to identify hidden patterns and design interventions that were unthinkable just a decade ago.

Genomic                                        Analysis
AI-driven genomics platforms can screen entire genomes for mutations, enabling personalized medicine tailored to each patient's unique genetic makeup.

Precision Diagnostics: With billions of base pairs to sift through in a single human genome, manual analysis is impractical. Machine learning algorithms rapidly compare a patient's genome against reference databases, flagging rare mutations that may predispose an individual to specific conditions such as Huntington's disease or inherited cancers.

Pharmacogenomics: Beyond diagnosing predispositions, AI can identify which therapies are most likely to succeed based on a person's genetic profile. For instance, some cancer drugs work best in patients with specific mutations. By aligning patient data with known genetic markers, clinicians can reduce trial-and-error treatments and customize regimens that optimize efficacy while minimizing side effects.

Population Genomics and Epidemiology: Large-scale genomic studies of entire populations are now feasible, helping public health officials track the spread of infectious diseases or identify emerging health risks. AI-powered analytics can reveal how a virus mutates across different regions or how environmental factors intersect with genetic vulnerabilities. This information is invaluable for crafting targeted interventions and resource allocation strategies, such as vaccine deployment or disease screening programs.

## Molecular Modeling

DeepMind's AlphaFold famously cracked the protein-folding challenge, accelerating the discovery of new therapeutics for conditions like Alzheimer's or cancer.

Protein Structure Prediction: For decades, biologists struggled with predicting the 3D shape of proteins from their amino acid sequences a key piece of understanding how proteins function in the body. AlphaFold's ability to predict structures with near-experimental accuracy has opened the floodgates for drug discovery, allowing researchers to model how new compounds might bind to a protein target without waiting for time-intensive lab crystallization techniques.

Rational Drug Design: With high-quality protein structures on hand, pharmaceutical companies employ AI to run virtual screening of thousands or even millions of chemical compounds. By simulating how each compound might fit into a protein's active site, they can rapidly shortlist the most promising drug candidates. This computational approach streamlines the early stages of drug development, potentially saving billions in research costs and years of trial-and-error.

Emerging Frontiers: AlphaFold-like systems aren't just limited to human proteins. They can model the proteins of pathogens (like bacteria or viruses) or even environmental organisms, guiding the creation of novel antimicrobial drugs or industrial enzymes. Future iterations may combine quantum computing with molecular modeling, potentially unveiling complex interactions at the subatomic level.

## Synthetic Biology

AI helps design synthetic organisms or genetic modifications to produce biofuels, biodegradable plastics, or disease-resistant crops.

Engineering Metabolic Pathways: Synthetic biology often involves reprogramming microbes (like yeast or bacteria) to manufacture substances such as insulin, bioethanol, or bioplastics. AI tools can predict how altering specific genes will impact a cell's metabolic pathways, speeding up strain optimization. For instance, a biofuel startup might use AI to identify the genetic tweaks that enable a microbe to convert agricultural waste into ethanol more efficiently.

Climate Resilience: In agriculture, AI-driven synthetic biology research has led to stress-tolerant crop varieties capable of withstanding drought, temperature fluctuations, or salinity. By pinpointing gene edits that strengthen root structures or improve photosynthetic efficiency, scientists aim to fortify the global food supply against climate-induced disruptions.

Ethical and Safety Considerations: Synthetic biology's power to reshape living organisms also raises regulatory and moral questions. If genetically modified microbes escape a lab setting, could they disrupt local ecosystems? Governments and international bodies are wrestling with biosafety protocols and dual-use concerns where technology intended for peaceful applications might be repurposed maliciously.

## Quantum Computing

Quantum computing represents a paradigm shift in how we process information. Whereas classical computers use bits (0s and 1s) to encode data, quantum machines employ quantum bits or "qubits," which can exist in multiple states simultaneously through superposition. This unique property, coupled with entanglement, could allow certain computations to be executed exponentially faster than on conventional hardware, triggering transformative changes in AI, cryptography, and more.

## Quantum Machine Learning

Theoretically, quantum circuits could process exponentially more states than classical computers, tackling tasks like large-scale optimization and simulating molecular interactions at a quantum level.

**Enhanced AI Algorithms:** Standard machine learning techniques might benefit from quantum speed-ups in training and inference, potentially enabling real-time analytics on massive datasets that today's supercomputers struggle to handle. Imagine a scenario where an AI processes global stock market data in milliseconds, identifying arbitrage opportunities or stability threats faster than any human trader.

Complex Problem Solving: Certain optimization problems like vehicle routing for logistics or advanced climate modeling could see dramatic gains from quantum approaches. For instance, a city's entire transportation grid could be optimized in near-real-time, factoring in traffic patterns, accidents, and weather to coordinate thousands of autonomous vehicles efficiently.

Cryptography
Post-quantum cryptography emerges to safeguard data against potential quantum hacks.

Breaking Classical Encryption: Many current encryption algorithms (like RSA) rely on the difficulty of factoring large numbers an obstacle quantum computer might overcome with algorithms such as Shor's. If quantum machines reach sufficient scale and stability, they could decrypt sensitive information government secrets, financial transactions, personal data that was once considered secure.

Developing Quantum-Safe Protocols: To preempt this threat, researchers and cybersecurity firms are crafting "quantum-resistant" encryption methods. These algorithms use mathematical problems thought to be resistant to quantum attacks, such as lattice-based cryptography. Governments and multinational corporations have begun trialing such protocols, anticipating a future where data protection demands quantum robustness.

Challenges
Quantum computing remains in its infancy affected by decoherence and error rates but intense research funding and corporate interest (Google, IBM, Intel) drive rapid advancements.

Fragility of Qubits: Qubits are highly susceptible to environmental noise; even tiny temperature fluctuations or electromagnetic disturbances can collapse their quantum state. As a result, quantum processors require specialized cryogenic environments and error-correcting techniques to maintain stability.

Scalability: While prototype systems can handle dozens or in rare cases, a couple of hundred qubits, practical quantum computers might need thousands or millions. Scaling up hardware, interconnects, and error-correction protocols remains a daunting engineering challenge.

Economic and Social Considerations: The cost of building and maintaining quantum hardware is astronomical, meaning early breakthroughs might be accessible primarily to tech giants and well-funded research labs. This raises concerns about "quantum inequality," where only certain actors hold the keys to computational leaps that could disrupt finance, national security, or scientific discovery.

## Aerospace, Robotics, and AI

Beyond terrestrial applications, AI is also reshaping how humanity explores and operates in outer space. From autonomous rovers on Martian soil to swarms of drones assisting in natural disaster relief on Earth, robotic systems guided by AI are expanding the boundaries of what's possible in extreme and challenging environments.

## Autonomous Space Exploration

Rovers on Mars (e.g., Perseverance) already employ AI for navigation, resource identification, and sample collection. Future missions might have more autonomy, learning on-the-fly to adapt to extraterrestrial terrain.

Navigational Autonomy: On a planet like Mars, communication delays can span several minutes. AI-driven systems help rovers make real-time decisions avoiding hazardous rock formations or sand traps without waiting for commands from Earth. This autonomy accelerates data collection and maximizes the mission's scientific returns.

Adaptive Science: In addition to obstacle avoidance, next-gen rovers may harness machine learning to identify geological formations that merit closer inspection. If a rover detects anomalies in soil composition, it can initiate its own sampling process or deploy mini-laboratories for on-site analysis, all while sending back highlights to mission control.

Drone Formations

Swarms of AI-driven drones could handle search-and-rescue operations during disasters or collect meteorological data in harsh environments.

Cooperative Algorithms: Instead of each drone operating in isolation, a swarm uses distributed intelligence to coordinate flight paths, avoid collisions, and divide tasks efficiently. These algorithms mimic certain insect behaviors like how bees share information about food sources allowing the swarm to cover large territories without duplicating effort.

Disaster Relief: After a hurricane or earthquake, drones can map collapsed structures and identify survivors needing urgent aid. Equipped with thermal imaging or lidar, they relay crucial information to rescue teams, even in inaccessible areas. This approach saves precious time, guiding ground crews to the spots most in need of help.

Space Manufacturing

In-orbit factories might use AI to manage complex processes, from microgravity-based 3D printing to robotics that construct satellites in space without human supervision.

Microgravity Advantages: Certain manufacturing processes like crystal growth or metal alloy formation can yield superior results in microgravity environments. AI systems can optimize these processes in real time, adjusting temperature controls or material feeds to produce higher-quality products than would be possible on Earth.

Self-Replicating Robotics: Some futurists envision fully automated satellite assembly lines orbiting Earth or stationed on the Moon. The robots would build and repair spacecraft, minimizing the need for risky human extravehicular activities. With AI overseeing production schedules and quality checks, these facilities could continuously adapt to new mission requirements or technological upgrades.

## Interdisciplinary Collaborations

The complexity of these frontiers biotechnology, quantum computing, aerospace, and more has driven computer scientists, biologists, sociologists, ethicists, and even anthropologists to work together in interdisciplinary teams. Where once AI research was confined to engineering labs, it now intersects with every domain that involves data, creativity, or complex problem solving.

Holistic AI Solutions: For instance, a biotechnology project may merge advanced ML algorithms (data scientists, AI researchers) with expertise in genetics (biologists) and patient advocacy groups (public health professionals). This collaborative ecosystem ensures that new treatments or diagnostic tools align with patient needs, ethical guidelines, and real-world constraints like insurance coverage or cultural acceptance.

Cultural and Ethical Insights: As AI shapes global society, social scientists and ethicists play key roles in anticipating unintended consequences. A robotics initiative might consult anthropologists to understand how communities will perceive or interact with automated systems, ensuring that technology integrates respectfully into local traditions and socioeconomic contexts.

Regulatory and Legal Input: Interdisciplinary teams often involve legal experts who stay abreast of emerging regulations. For instance, quantum computing might raise patent issues, while biotech breakthroughs prompt debates about gene-editing patents or biosafety standards. Lawyers help translate technical innovations into policy recommendations that balance public interest with private sector incentives.

## Brain-Computer Interfaces and Next-Gen AI: Neuralink and OpenAI's o3

One of the most dramatic frontiers emerging is the convergence of neural interfacing and advanced AI. By blending the capabilities of implantable devices and sophisticated machine intelligence, researchers envision a world where humans can interact with digital systems through thought alone, or even augment their own cognitive abilities in real time.

Neuralink's Trajectory: Founded by Elon Musk, Neuralink is pioneering implantable brain-computer interfaces (BCIs) that integrate seamlessly with the human brain.

Restoring Lost Function: In 2024, Neuralink initiated human trials demonstrating a paralyzed patient controlling a computer with thought alone a milestone that showcased the promise of BCIs in restoring lost function. The patient could manipulate on-screen cursors, type messages, and even play simple games without any muscular input.

Augmenting Human Capabilities: By mid-year, a second participant demonstrated unprecedented mastery of design software using only brain signals, highlighting the potential for BCIs to augment human creativity in fields like art, architecture, and science. Meanwhile, researchers advanced prosthetics, unveiling brain-controlled robotic limbs that move with fluidity almost indistinguishable from natural limbs.

Future Visions: Neuralink's roadmap includes enhanced memory, accelerated learning, and the possibility of bridging sensory deficits such as restoring sight or hearing through direct neural stimulation. While these goals inspire awe, they also compel rigorous safety testing, psychological counseling, and careful consideration of potential social disparities.

OpenAI's o3 Model: In parallel, OpenAI's newest breakthrough its "o3 model" pushes AI reasoning closer to human-level cognition.

Cognitive-Inspired Architecture: Unlike earlier models reliant primarily on pattern recognition, o3 features a cognitive-inspired framework designed to "think" and solve problems much like humans do only faster. This involves internal "reasoning loops" that allow the AI to check its own logic and refine conclusions before responding.

Applications: Healthcare providers began using o3 to generate personalized treatment strategies for complex conditions, while educators tapped into o3's capacity to craft individualized learning modules. In software development, engineers rely on o3's ability to not just write code, but to self-evaluate, debug, and propose optimizations in a fraction of the time it would take human programmers.

Open-Source Variants: OpenAI, meanwhile, pursues open-source variants of o3 (such as "o3-Mini") to democratize AI. This ensures that small businesses, academic labs, and even hobbyist developers can harness advanced reasoning tools without requiring supercomputer-level resources.

## The Dawn of "Hybrid Intelligence"

When Neuralink's BCI meets OpenAI's o3, the line between human and machine blurs further. Imagine a user mentally formulating a request, and o3 instantly interpreting it to solve a problem, compose music, or optimize a design. This real-time loop opens the door to "hybrid intelligence," a scenario where human intuition merges with AI's computational might. Enthusiasts argue that this synergy could herald a new evolutionary step freeing us from physical and cognitive constraints. They foresee professions where doctors, designers, and scientists wield supercharged mental toolkits, drastically cutting research and development timelines.

However, critics raise valid questions about autonomy, security, and equity. If AI can access and interpret neural signals, who safeguards that data, and under what regulations? Could hackers exploit neural implants for malicious ends, controlling or monitoring a user's thoughts? And what about the risk of societal stratification between those who can afford cognitive enhancements and those who cannot?

Data Ownership and Mental Privacy: Brain data could become the ultimate personal identifier, revealing a person's emotional states, preferences, and vulnerabilities at an unprecedented granularity. Experts argue that new legal frameworks perhaps akin to GDPR will be needed for "mind data," including the right to erase or anonymize neural records.

Potential Surveillance: Advanced AI might manipulate or predict human behavior at an alarming scale, especially if corporations or governments gain real-time access to our mental states. Privacy advocates fear a future in which mental intrusions become the norm, requiring robust encryption and strict usage policies for neural data.

Empowerment vs. Control: Ensuring that these innovations remain tools of empowerment rather than instruments of control will require interdisciplinary oversight, public engagement, and a global commitment to equitable access. Governments might stipulate those neural enhancements be regulated as medical devices, limiting commercialization until safety and ethical standards are met. Conversely, militaries might explore "augmented soldiers," fast-tracking classified research that accelerates the technology's development in less transparent ways.

Ultimately, these developments hint at a future in which "downloading" knowledge, controlling everyday devices via thought, and interacting with AI on an intimate cognitive level may become commonplace. In some optimistic visions, BCIs and advanced AI could democratize expertise, usher in new creative renaissances, and help solve crises ranging from climate change to public health. Yet the social, ethical, and political dimensions are equally profound. Who decides which features get prioritized? How do we protect individual freedoms in a world where minds can be linked to digital networks?

The challenge and promise of next-gen AI and neural interfacing is thus twofold. On one hand, these technologies can significantly expand human potential, enhancing our intellect and physical capabilities in ways once relegated to science fiction. On the other, they intensify age-old dilemmas about power, identity, and morality. Navigating this landscape responsibly will demand not just technological brilliance, but also compassionate governance, inclusive design, and global cooperation.

As we stand on the threshold of brain-computer integration and AI that rivals (or surpasses) human cognition, the conversation must extend beyond technical feasibility into the realm of collective values. By engaging stakeholders from all backgrounds policymakers, ethicists, everyday citizens, and the private sector society can strive to ensure that the next leaps in biotechnology, quantum computing, and neural AI become stepping stones to a more equitable and enlightened future, rather than catalysts for new forms of inequality or exploitation.

# Javad Yahaghi

# Part 5: Empowering the Individual from Consumer to Creator

# Chapter 16: Your Role in the AI Revolution

## Becoming AI-Literate

AI is rapidly reshaping industries, social structures, and even personal life choices. Despite this widespread influence, many people still feel uncertain about how AI works or how to separate genuine breakthroughs from overblown promises. Cultivating AI literacy knowing the basics of machine learning, data privacy, and algorithmic decision-making empowers individuals to navigate this evolving landscape with confidence. Below are some practical ways to increase your AI fluency, whether you're a student just getting started or a professional seeking new skills.

## Online Learning Resources

Platforms like Coursera, edX, fast.ai, and Kaggle: These websites offer a range of courses from introductory machine learning to more advanced topics like deep neural networks or data ethics. Some are self-paced and free to audit, making them accessible to almost everyone with an internet connection.

**Demystifying AI Concepts:** If math-heavy tutorials feel overwhelming, look for courses that emphasize conceptual understanding rather than code. Many platforms provide intuitive explanations and real-world examples, illustrating how supervised learning, unsupervised learning, and reinforcement learning operate in practical settings.

**Spotting AI Hype vs. Reality:** By grasping core ideas like how models learn from data and the importance of avoiding bias you can better judge a company's claims about "groundbreaking AI." When you see headlines announcing an "AI that reads minds" or "revolutionizes healthcare in minutes," you'll know enough to ask critical questions about data sources, methodology, and peer review.

## Hands-On Experimentation

**Beginner-Friendly Tools:** Platforms such as Google's Teachable Machine or Create ML for Apple ecosystems let non-experts build simple ML models with drag-and-drop interfaces. They typically guide you through collecting small datasets like images of different objects or short audio clips and training a model to recognize them.

**Immediate Feedback Loop:** Because these tools handle the underlying code, you see results quickly. For instance, after uploading a few photos of your pet dog, the platform can generate a mini classifier that identifies similar images. This immediate feedback provides a sense of how data quantity and quality affect model accuracy.

**Scaling Up:** Once comfortable with these entry-level experiences, you might explore more sophisticated frameworks (TensorFlow, PyTorch) or specialized libraries (OpenCV for computer vision, NLTK for natural language processing). Over time, you'll gain a deeper understanding of hyperparameters, model architectures, and the iterative process of improving AI applications.

## Engaging with AI in the Workplace

Whether you're a teacher, lawyer, marketer, or mechanic, integrating AI into daily tasks can significantly boost efficiency and open doors to new opportunities. Yet, true engagement goes beyond installing a chatbot or analyzing a data report. It involves a proactive mindset, a willingness to learn, and a commitment to using AI in ways that respect both ethical principles and human creativity.

## Upskilling

Streamlining Repetitive Tasks: In many professions, repetitive duties like data entry, scheduling, or preliminary research consume valuable time. By automating these tasks with AI tools, professionals can free up their schedules for strategic thinking, client engagement, or creative work. For example, a lawyer might use AI to sift through case law, focusing personal attention on shaping a winning argument.

**Data-Driven Insights:** Marketers can glean detailed customer patterns using machine learning analytics, tailoring campaigns more effectively. Teachers can use adaptive learning platforms that track student progress, customizing lesson plans based on individual needs. Even mechanics could adopt AI-powered diagnostic systems that scan vehicle performance data in real time, predicting potential failures and reducing downtime for customers.

**Continuous Learning and Adaptation:** As AI evolves, so do the skill sets required to leverage it. Workshops, online training modules, and internal knowledge-sharing events can help you remain current. Recognizing that AI is not a one-time introduction but an ongoing innovation can position you and your organization for sustained success.

## Champion Responsible AI

**Fairness Checks and Privacy Safeguards:** Encourage your organization to implement fairness audits, particularly if AI systems handle sensitive data like hiring, credit scoring, or health diagnoses. Ask whether the model has been tested across diverse populations or if anonymization techniques are in place to protect user identities.

**Transparent Documentation:** Documenting AI processes, from data collection to model deployment, ensures accountability. Advocate for clear data governance policies and plain-language explanations of how algorithms make decisions especially if they significantly impact employees, customers, or the public.

Ethical Advocacy: Become a spokesperson for ethical AI within your team. Highlight success stories where responsible AI led to tangible benefits (e.g., unbiased hiring practices, safer workplaces), and call attention to risks when corners are cut. This proactive stance can help steer your organization away from reputational crises or regulatory penalties down the line.

## Civic and Community Involvement

AI's reach isn't limited to the corporate world or specialized labs it increasingly shapes public policy, urban planning, and social services. By engaging at a community or governmental level, individuals can influence how AI is adopted in crucial domains like education, law enforcement, and environmental management. Your voice matters, whether you're raising ethical concerns, proposing innovative solutions, or participating in local debates about data privacy.

## Public Consultations

**Commenting on AI Regulations:** Many city councils, state agencies, or national governments invite public feedback on proposed AI legislation such as facial recognition bans or data privacy acts. By attending hearings or submitting written comments, you can draw attention to overlooked issues, suggest evidence-based improvements, and ensure that policy reflects diverse perspectives.

**Town Halls and Community Forums:** Keep an eye out for local events where policymakers discuss new AI initiatives (like "smart city" programs or predictive policing). Showing up and encouraging neighbors or colleagues to join can bolster transparency and accountability, as officials who hear well-informed questions may reconsider potentially flawed designs.

## Grassroots Data Projects

**Citizen Science Initiatives:** Mapping local environmental data, cataloging historical archives, or crowdsourcing birdwatching logs are just a few ways communities contribute to open-source datasets. These records can fuel AI models that help monitor air quality, preserve cultural heritage, or predict wildlife migrations.

**Collaborative Platforms:** Websites like Kaggle or GitHub host public data challenges, where volunteers can collectively build or improve AI solutions for social good. By contributing data or code or simply offering domain expertise you help create resources that benefit nonprofits, researchers, and citizen advocacy groups.

**Local Partnerships:** Organize data collection drives or hackathons in collaboration with schools, libraries, or community centers. For instance, a neighborhood might pool smartphone photos of potholes or broken streetlights, giving the city a clearer picture of infrastructure needs. AI developers can train models to prioritize repairs, ensuring that municipal authorities act on reliable, up-to-date information.

## Activism and Whistleblowing

**Highlighting Unethical Uses:** Workers inside big tech companies have raised concerns about surveillance projects or manipulative recommender systems. By publicly disclosing questionable practices, whistleblowers can prompt internal reforms or spark public debates that lead to tighter regulations.

**Support Networks:** Whistleblowing often carries professional risks, including job loss or legal repercussions. Organizations like the Government Accountability Project (GAP) or the Whistleblower Aid group provide legal guidance and moral support. Unionizing tech workers is another strategy to protect collective rights and uphold ethical standards.

**Global Solidarity:** In an interconnected world, AI practices in one country can quickly influence others. Activists can coordinate internationally, sharing legal precedents, best practices, and strategies for pressuring multinational corporations. By forging cross-border alliances, civic movements can shape AI's trajectory on a scale that transcends national boundaries.

In essence, AI literacy and active engagement aren't reserved for tech specialists. Whether you're learning the basics online, introducing AI tools at your workplace, or raising your voice in civic forums, each action contributes to a broader culture of responsible innovation. The decisions we make now on transparency, fairness, and user empowerment will guide AI's impact for years to come, influencing everything from individual livelihoods to the health of democratic institutions.

By becoming AI-literate, championing ethical applications, and participating in public discourse, you help ensure that AI develops as a genuinely transformative force, one that uplifts communities rather than undermining them.

# Chapter 17: Augmenting Human Potential

## Co-Creation and Collaboration

AI can handle large-scale data analysis, pattern recognition, and repeated tasks. Humans excel at ethical judgment, empathy, strategic thinking, and cross-domain creativity. The power of modern technology lies not in replacing human capabilities, but in weaving them together with AI's strengths to achieve novel outcomes that surpass what either could accomplish alone. Below, we delve deeper into how different sectors harness this synergy:

## Creative Professions

Designers, composers, and authors use AI as a springboard for ideas generating initial drafts or offering style suggestions while the human retains final editorial control.

Idea Generation and Experimentation: In fields like fashion design, AI can propose new color palettes or material combinations by analyzing current trends and historical data. The designer then chooses which suggestions to refine, ensuring the final creation reflects their unique vision. Similarly, in literary settings, an author might prompt an AI to propose multiple plot twists or character arcs, selecting only those that resonate with the story's themes.

Emergent Aesthetics: By analyzing millions of images or soundscapes, AI can uncover aesthetic patterns that transcend cultural boundaries. This can inspire composers to blend genres in unexpected ways or prompt graphic artists to experiment with color schemes they might not have encountered otherwise. However, it's ultimately the artist's intuition, emotional insight, and storytelling ability that transform a raw AI-suggested concept into an evocative piece.

Balancing Control and Spontaneity: One of the biggest debates in AI-augmented art is how much autonomy to grant the machine. Some creators treat AI purely as a tool for instance, a quicker way to draft background elements or refine minute details. Others embrace co-creation fully, letting the AI's randomness spark accidental brilliance. In both cases, the human remains the ultimate arbiter of quality, taste, and meaning.

## Scientific Discovery

Researchers rely on AI for screening massive datasets (astronomical observations, genomic sequences), then apply human intuition to interpret novel patterns and propose theories.

Data Overload Solutions: In astrophysics, for example, telescopes capture terabytes of information daily. Manually combing through these datasets for hints of exoplanets or supernovae is impractical. AI not only flags potential anomalies but also classifies them by likelihood of scientific interest. From there, researchers apply domain expertise to confirm or challenge the AI's leads.

Hypothesis Generation: In biology, AI might suggest connections between genetic markers and disease symptoms that had never been considered. Scientists then design lab experiments to test those connections, refining the theoretical framework as they go. This interplay accelerates breakthroughs, whether identifying new drug targets or understanding the mechanics of aging.

Redefining Peer Collaboration: With AI handling the grunt work of data parsing, scientists can focus on higher-level discussions and cross-disciplinary exchanges. A climatologist and a microbiologist might co-author a paper on how rising sea temperatures affect plankton populations, all backed by an AI's real-time analysis of oceanic satellite data.

## Social Services

AI chatbots offer first-level counseling, scheduling, or administrative support, freeing human counselors to focus on high-empathy interactions.

Augmenting Public Health: Community clinics use chatbots for basic triage asking patients about symptoms and directing them to relevant services. This reduces wait times and operational costs, but human physicians still oversee final diagnoses and care plans.

Streamlined Case Management: In social work, AI can track clients' appointments, remind them of required documentation, and even offer 24/7 crisis lines for urgent situations. Social workers then have more bandwidth to build trust and provide deeper emotional support, rather than juggling administrative tasks.

Long-Term Relationship Building: Crucially, there's still no replacement for genuine human empathy. A chatbot might recognize anxiety in a user's text messages, but a seasoned counselor can perceive nuances in tone or context that an algorithm might miss. The result is a complementary approach: the AI ensures immediate accessibility, while the professional nurtures therapeutic bonds that help clients navigate complex emotional or socioeconomic challenges.

## Rethinking Professional Roles

As AI systems become entrenched in day-to-day workflows, professions evolve. Some traditional roles may dissolve, but others emerge, reflecting the growing need for new skills and ethical frameworks. Society stands at a crossroads: either we usher in an inclusive transformation where workers receive support to adapt or we risk deepening inequalities as certain groups are left behind.

# New Job Categories

"Prompt Engineers" refine textual inputs to AI systems for optimal outputs. Their role is to craft the context and instructions that guide an AI's problem-solving. This skill has gained prominence in creative and analytical fields alike, from marketing copy to data analytics.

"AI Ethicists" assess social impacts. These professionals delve into how algorithmic decisions might affect vulnerable populations, scrutinize data governance policies, and shape internal codes of conduct. Think of them as modern-day compliance officers, but with a focus on fairness, bias mitigation, and broader societal well-being.

Domain-Specific Hybrids: In medicine, for instance, "clinical data scientists" merge knowledge of patient care with AI modeling to create diagnostic tools or personalized treatment plans. Legal-tech consultants use AI for contract reviews, while also ensuring new software aligns with regulatory and privacy norms.

# Reskilling Pathways

Governmental Investment: As automation reshapes industries, policymakers can fund large-scale education initiatives scholarships for AI-related courses, subsidized coding bootcamps, or skill certification programs. This proactive approach helps displaced workers pivot faster to emerging roles, whether that's data wrangling or AI project management.

Corporate Responsibility: Forward-thinking companies recognize that investing in employee training is more sustainable than cyclical layoffs. Internal workshops on machine learning basics or data visualization can spur an innovation culture. Some firms even partner with universities to create custom curricula, ensuring graduates master the exact competencies needed in the field.

Life-Long Learning Mindset: On an individual level, the rapid pace of AI advances means formal education won't be a one-and-done. E-learning platforms, mentorship circles, and micro-credential programs will likely become a staple. People who embrace continuous upskilling staying curious about new technologies stand to flourish in AI-augmented environments.

## Emotional and Ethical Labor

Human Strengths: Counseling, moral deliberation, spiritual guidance, and creative storytelling tap into uniquely human capacities like empathy and cultural sensitivity. Even as AI encroaches on logical or data-heavy tasks, these spheres remain distinctly human.

Elevating Purpose: Freed from repetitive chores, many professionals can devote more energy to big-picture strategy, interpersonal connection, or fostering workplace community roles that a machine can't replicate. Employers that acknowledge and reward these soft skills may retain a more engaged, holistic workforce.

Javad Yahaghi

Guardrails Against AI Overreach: Humans must remain the final arbiters of decisions that carry deep ethical weight. This is especially vital in healthcare (life-and-death choices), education (child development), or criminal justice (sentencing and rehabilitation). Even the most advanced algorithmic tools need oversight by individuals trained to consider moral, cultural, and psychological nuances.

## Personal Fulfillment in an AI-Driven World

As routine tasks automate, humans might find greater time for reflection, collaboration, and creative pursuits. The extent to which societies leverage this newfound freedom hinges on how leaders allocate resources and plan for systemic shifts. Below are several potential outcomes.

## Renaissance of Ingenuity

Creative Exploration: With mundane work delegated to AI, aspiring musicians, writers, or inventors can indulge their creative urges more freely. Grants, communal workspaces, and mentorship programs can further nurture these talents, spurring cultural renaissances.

Collaborative Science and Art: Interdisciplinary labs could pair historians with data scientists to reinterpret archeological records using AI, or choreographers with technologists to craft performances blending holograms and live dance. Such collective endeavors push the boundaries of human expression, fueled by ample time and cutting-edge AI support.

## Risks of Social Division

Ignoring Workforce Displacement: If policy frameworks fail to provide adequate support like job retraining, income assistance, or mental health resources laid-off workers may experience economic hardship or social resentment. Large-scale unemployment could spark civil unrest, undermining political stability.

Ethical Oversight Lapses: Without robust regulations, unscrupulous corporations might exploit AI to reduce labor costs without reinvesting in worker welfare. Automated hiring, for instance, could quietly perpetuate bias if no one's monitoring fairness metrics. Overreliance on predictive policing could entrench discriminatory practices, eroding public trust.

Uneven Access to AI Benefits: Geographic and socioeconomic disparities might deepen if only privileged communities have reliable tech infrastructure. Urban centers might thrive with advanced AI in schools and hospitals, while rural or underfunded areas lag. Bridging this "digital divide" requires strategic public investment and committed local leadership.

Javad Yahaghi

## Reshaping Educational Systems and Moral Frameworks

Evolving Curricula: Traditional rote learning becomes less critical when data retrieval is instant and automated. Instead, schools may emphasize soft skills like collaboration, critical thinking, and emotional intelligence skill sets that AI cannot readily replicate. Hands-on projects, community service, and interdisciplinary problem-solving modules help students adapt to a world where learning extends beyond memorizing facts.

Work Cultures and Values: As machines take on more routine tasks, companies might adopt flatter hierarchies or flexible work structures. Employees, rather than being cogs in a production line, could become "innovation stewards," scanning for improvement opportunities or forging cross-departmental collaborations.

Moral Considerations: With greater leisure and creative autonomy, individuals and communities might revisit age-old debates on purpose, ethics, and spirituality. If machines can handle logistics and computation, humans can collectively explore philosophical questions about identity, empathy, and the pursuit of wisdom, potentially leading to a more introspective and values-oriented society.

## Beyond Restoration: Neuralink's and o3's Path to Cognitive Enhancement

The synergy of Neuralink and o3 moves us from merely restoring lost functions like controlling prosthetics or aiding the visually impaired toward augmenting human cognition itself. This leap offers tantalizing prospects for expanding knowledge, creativity, and problem-solving abilities, but it also heightens concerns about who gets access and how these enhancements might reshape social norms.

## "Downloading" Knowledge

Accelerated Mastery: Imagine learning a new language or rapidly mastering complex software through a direct brain-to-AI interface. The friction of repetitive study, skill drills, or memorization could evaporate, replaced by near-instant data transfers. Instead of spending years mastering 3D modeling, an architect might install the relevant knowledge modules within days.

Rethinking Traditional Learning Systems: Schools and universities would have to adapt. Do we still need multi-year programs when entire curriculums could be absorbed more quickly? Does this risk diminishing the cultural and social experiences tied to conventional education? Some academics worry about losing essential critical thinking or personal development that emerges from time-intensive study.

Mental Health and Cognitive Overload: Plugging an AI directly into human thought processes could blur the line between personal identity and external knowledge. Would individuals feel overwhelmed or lose a sense of self if they "download" new competencies daily? Psychologists and ethicists stress the need to measure cognitive load and emotional well-being, ensuring that empowerment does not spiral into digital dependency or identity crises.

## Augmenting Creativity and Problem-Solving

Global Innovation Boom: Enthusiasts view this future as a renaissance of human potential, accelerating breakthroughs across science, art, and philosophy. Scientists might formulate new theories at a rate never before seen, while artists collaborate with AI in real time, forging multimedia experiences that push creative boundaries.

Equity Gaps: Yet skeptics point to the risk that only an elite few might afford cognitive boosts, thereby widening social divides. If certain regions or demographic groups gain access first, disparities in productivity and economic influence could balloon. This raises urgent policy questions: Should governments subsidize neural interfacing for the public good, or will private markets control distribution?

Regulatory and Ethical Challenges: Deciding who qualifies for enhancements, under what conditions, and how to regulate their use is no small feat. Some fear a "race to the top" where competitive pressures force professionals in high-stakes industries like finance or healthcare to adopt neural implants just to keep pace. Others caution that mandatory cognitive augmentation would infringe on basic human rights, especially if coerced by employers or state agencies.

## Shaping a Framework for Inclusive Progress

Policymakers, Ethicists, and Citizens must collaborate to ensure that cognitive enhancement respects human dignity, fosters inclusive progress, and preserves the richness of diverse learning paths. Community debates might revolve around caps on enhancement levels, mandatory mental health checks, or data transparency laws (e.g., prohibiting the resale of neural data).

Safeguarding Individual Autonomy: Even if technology allows advanced AI to interface seamlessly with the human brain, consenting to this connection should remain a personal choice rather than a corporate or governmental mandate. Transparent guidelines on data usage, ownership, and liability will be crucial to maintaining trust in these systems.

Global Coordination: Given that biotech and AI cross national borders, international treaties or agreements might be necessary to prevent uneven distribution or unethical exploitation. Bodies like the United Nations or the World Health Organization could facilitate discussions on universal standards, similar to how they address global health crises or nuclear proliferation.

In conclusion, the co-creation model between AI and human expertise underscores the power of collaboration both in the workplace and in broader societal structures. As we refine professional roles and adapt our moral frameworks to an AI-driven landscape, we gain the freedom to pursue deeper creative, scientific, and empathetic endeavors. At the pinnacle of this transformation, technologies like Neuralink and the o3 AI model offer a glimpse of what true human-machine fusion might look like a future where advanced intelligence supports rather than supplants human value systems.

The key to unlocking AI's promise, however, lies in thoughtful governance, inclusive access, and an unwavering commitment to upholding human dignity. By shaping policies that prioritize education, equity, and ethical deployment, we can harness AI to amplify human potential while preserving the rich tapestry of personal autonomy and social cohesion that makes us fundamentally human.

# Chapter 18: From Insight to Action Tools and Next Steps for Engagement

## Practical Guides and Resources

While AI may seem intimidating at first, gaining hands-on skills and understanding the broader ethical landscape is increasingly within reach. Many organizations, from grassroots community groups to multinational tech firms, are providing resources and platforms that make it easier for individuals to learn, collaborate, and influence AI's development. Below are some practical ways to get involved, whether you're a curious parent, an educator, an industry professional, or just someone eager to contribute to the responsible growth of AI.

## AI Literacy for All

## Code.org Machine Learning Modules

K–12 Accessibility: Traditionally focused on teaching basic coding, Code.org now offers machine learning modules aimed at younger learners. These modules introduce core concepts like how algorithms recognize patterns or classify images through interactive puzzles.

Classroom Integration: Many schools already use Code.org's curriculum to teach fundamental programming skills. Adding AI lessons helps demystify machine learning at an early age. Instead of waiting until college or a specialized tech bootcamp, children can grow up seeing AI as just another tool in their problem-solving toolkit.

Equitable Outreach: By reaching students from various backgrounds, these modules contribute to a more diverse AI talent pipeline. Early exposure boosts confidence and can spark an interest in STEM, particularly among underrepresented groups.

## Adult Learning Opportunities

DataCamp, Udacity, and Khan Academy: These platforms expand adult learning, offering courses from fundamental statistics to practical data science projects. Some focus on foundational math and programming, while others dive deeper into neural networks, computer vision, or NLP.

Flexible Formats: Busy adults often have limited time, so many online courses are self-paced. Bite-sized lessons, quizzes, and project-based tasks make it feasible to learn in short increments.

Pathway to Professional Development: Completing a sequence of courses can bolster your résumé, helping you pivot into an AI-related job or enhance your current role with data-driven insights. Many learners use these credentials to negotiate better positions or initiate AI projects in their organizations.

## Open-Source Communities

## GitHub

Thousands of AI Libraries: GitHub hosts repositories for everything from basic machine learning utilities to complex frameworks like PyTorch or TensorFlow. Contributing to open-source AI libraries by writing documentation, fixing bugs, or adding features offers hands-on experience and helps you learn from seasoned engineers.

Real-World Problems: Many open-source projects tackle challenges like medical imaging, climate modeling, or language translation. Getting involved means you're not just practicing AI in isolation; you're helping solve tangible issues that matter to communities worldwide.

Portfolio Building: Submitting pull requests and maintaining your own repositories can establish a public record of your work. Potential employers, collaborators, or mentors can view your code quality and creative problem-solving skills.

## Hugging Face

Transformers and NLP: Known for its user-friendly interface, Hugging Face simplifies access to state-of-the-art language models like BERT, GPT, and others. You can easily train, fine-tune, or deploy models for tasks ranging from sentiment analysis to text generation.

Collaborative Space: Hugging Face's "Model Hub" and "Spaces" invite users to share pre-trained models and interactive demos. This transparency encourages peer learning developers can critique each other's approaches, optimize performance, and even create custom solutions for specialized languages or domains.

## Ethical Frameworks & Audits

## "Ethics by Design" Committee

Proactive Approach: If you work in an organization deploying AI systems whether in finance, healthcare, education, or e-commerce champion the formation of an internal ethics committee. This group would regularly audit data sources, review model outcomes for bias, and ensure compliance with regulations such as GDPR or the EU AI Act.

Interdisciplinary Membership: Ideally, the committee includes data scientists, legal experts, diversity advocates, and domain specialists who can offer varied perspectives on potential harms.

Employee Engagement: Invite input from colleagues at all levels. A customer support representative, for example, might notice AI-driven recommendation systems that inadvertently disadvantage certain customer segments.

## Bias Assessment Tools

IBM's AI Fairness 360: A toolkit featuring metrics and algorithms to detect and mitigate bias in datasets or model predictions.

Google's Model Card Toolkit: Encourages systematic documentation of a model's purpose, performance, and limitations across different demographic groups.

Regular Monitoring: Conduct these audits not just at launch but periodically, as real-world data shifts over time. If a system starts showing signs of biased outcomes, the committee can intervene retraining the model, diversifying the dataset, or imposing additional review steps before deployment.

## Local and Global Policy Engagement

## Writing to Representatives and Public Forums

Influencing Legislation: Policymakers draft AI regulations that impact everything from privacy rights to national security. Even a short letter or email expressing concerns or supporting certain bills can guide political priorities.

Public Forums: Many localities host town halls or community meetings on AI-related topics, such as facial recognition in policing or data-sharing agreements with tech companies. Attending, asking questions, and sharing personal experiences can shape the debate.

## Monitoring Legislation

EU AI Act Updates: The European Union's evolving risk-based framework for AI can set global precedents. Keeping an eye on amendments or newly proposed articles helps you anticipate compliance requirements.

U.S. Federal Guidelines: Beyond the Blueprint for an AI Bill of Rights, various legislative proposals covering areas like facial recognition, automated decision-making, and biometric data are under discussion. Being informed allows you to support or challenge these initiatives effectively.

Data Governance in Asia: Countries like China, Singapore, and India have their own data regulations and AI strategies. Tracking these developments illuminates global trends and fosters cross-cultural collaboration on ethical practices.

## Civic Tech Groups or NGOs

Equitable and Responsible AI: Organizations like the Algorithmic Justice League or Access Now advocate for transparency, accountability, and human rights in the digital realm. You can volunteer expertise, donate, or simply amplify their research via social media.

Local Chapters: Search for meetups or hackathons in your area organized by civic tech enthusiasts. These events often focus on local issues (e.g., improving public transportation, analyzing municipal budgets with open data) and welcome diverse participants from developers to concerned citizens.

## Scaling Individual Efforts into Collective Impact

While your personal actions learning a new skill, voting on local AI regulations, contributing to open-source may seem small, the cumulative effect can be profound. Tech history is filled with grassroots movements that reshaped the landscape, from the rise of Linux (which began as a hobby project) to open data initiatives that transformed public sector transparency. In AI, a similarly collective effort can steer the technology toward equitable, inclusive outcomes.

## Grassroots Movements

Open-Source Software Parallels: Just as open-source projects democratized software development, concerted citizen engagement can democratize AI. People from all walks of life journalists, activists, educators, programmers collaborate to create tools, guidelines, or data sets that benefit everyone.

Cascading Effects: Your small contribution testing a bias detection plugin, writing a blog post about an AI use-case in your community might inspire others to do the same. Over time, these individual initiatives add up to widespread changes in the tech industry's culture and priorities.

# From Passive Consumer to Active Co-Creator

Shifting Mindsets: Rather than merely adapting to AI-driven products (like new chatbots or personalized recommendations), you can question their design, propose improvements, or even build your own. This shift from passive consumption to active creation is central to ensuring AI reflects diverse needs and values.

Collaborative Ownership: Publicly funded or community-driven AI projects spread ownership beyond corporate boardrooms. Through crowdfunding, academic partnerships, or government grants, communities can develop AI solutions tailored to local challenges healthcare in rural settings, language tools for minority dialects, or environmental conservation in sensitive ecosystems.

Long-Term Cultural Impact: When civic-minded people regularly engage with AI through local policy debates, neighborhood data projects, or even viral social media campaigns it sets a precedent for future generations. Young people growing up in these communities see AI not as a mysterious force to be feared but as a collective resource, shaped by and for the public good.

By participating, you transform from a passive consumer of AI into an active co-creator of its future. Each course you take, hackathon you join, or public forum you attend contributes to the broader tapestry of responsible AI development. In a field as dynamic and influential as artificial intelligence, even small, localized actions can help safeguard democratic values, promote social equity, and ensure that technological progress remains centered on human well-being.

# Conclusion

## Reflecting on the Journey and Embracing Our Shared Responsibility

Over the course of this book, we have traveled from the fundamentals of AI machine learning, deep learning, neural networks to its deep footprints in industries like healthcare, education, finance, transportation, and creative expression. We have engaged with the tough moral and philosophical questions AI raises: its potential for bias, risks of misuse, concentration of power, and even the mystery of consciousness. We have surveyed possible future scenarios, from "smart city" infrastructure to global sustainability missions, from quantum computing breakthroughs to interplanetary robotics. And crucially, we have recognized that every individual has a role in steering this technology, whether through technical contributions, policy advocacy, or thoughtful use of AI tools in everyday life.

Yet if we step back, it's clear that our journey with AI is only beginning. Every innovation we've explored from medical diagnostics systems that save lives by catching disease early, to neural networks that compose symphonies remains a stepping stone on a road that extends far beyond our current horizon. Technologies like advanced brain-computer interfaces or self-organizing drone swarms still appear futuristic to some, but tomorrow they may become as ubiquitous as the smartphone.

With each breakthrough comes new ethical dilemmas, new governance challenges, and new possibilities for human flourishing. This tension between potential and peril defines the dynamic nature of AI.

## AI as a Shaping Force, not a Fixed Destiny

One of the most important insights from our collective exploration is that AI is not destiny it is a shaping force that reflects our collective decisions about data, governance, ethics, and societal priorities. In other words, AI does not develop in a vacuum; it emerges within complex human systems. Corporations invest in certain research directions, governments impose (or fail to impose) regulations, and communities voice support or concern about how these technologies invade or enhance their daily lives. These factors ultimately shape which AI applications flourish, which remain on the periphery, and which are deemed unacceptable or too harmful to pursue.

## Values and Ethics as Guiding Principles

AI technologies can amplify existing inequalities if they are deployed recklessly, or they can serve as powerful tools for social good if guided by principles like equity and transparency. This dichotomy underpins the urgency to embed ethical guidelines into every phase of AI development and deployment. Whether we're talking about data collection for predictive policing or using generative models in healthcare, the underlying values we choose to prioritize such as fairness or autonomy will shape outcomes as surely as any line of code.

## Democracy and Collective Participation

Far from being the domain of experts alone, AI invites community engagement. Local councils debate the use of facial recognition; national legislatures propose bills to limit or encourage certain types of automation; international bodies attempt to harmonize standards. In each of these arenas, citizens' voices matter. When individuals volunteer in grassroots data projects, attend town hall meetings on tech policy, or advocate for fairness audits in their workplaces, they exert real influence on how AI evolves.

## Innovation vs. Regulation

The conversation often frames AI innovation and regulation as adversaries, but a balanced approach recognizes that well-crafted regulations can spur rather than stifle creativity. Clear, fair rules help smaller players compete on a level field with tech giants, ensure that communities consent to how their data is used, and reduce the risk of catastrophic outcomes that could erode public trust. Smart governance is not about hindering progress; it's about channeling the transformative power of AI in ways that align with shared human values.

## The Power and Perils of Acceleration

As innovation accelerates, the stakes multiply. Advanced AI can help solve climate change, accelerate vaccine development, or revolutionize education, but it can also enable invasive surveillance, entrench power imbalances, and erode civil liberties if left unchecked. It is precisely this dual potential that makes public oversight, ethical frameworks, and transparent governance so critical.

## Concentration of Power

We've seen how AI can centralize resources in the hands of a few. Corporations and governments with access to massive datasets and computational infrastructures often hold disproportionate sway over how AI is funded and deployed. This concentration raises questions about monopolies, privacy breaches, and even democratic integrity if deepfake technologies or micro-targeted propaganda undermine rational discourse.

## Existential Risks

The arrival of Neuralink's advanced brain-computer interfaces and OpenAI's o3 model reminds us that the frontier is always evolving. While these breakthroughs could spark a new age of creativity and cognitive liberation, they also unearth deeper philosophical and existential questions. If machines can mimic or surpass human cognition, how do we ensure their goals align with ours? How do we retain our sense of autonomy, privacy, or identity when thoughts themselves can be digitized or influenced by AI-driven feedback loops? Although the science fiction scenarios of rogue superintelligence may seem distant, the rate of change suggests we should proactively shape these pathways rather than wait to react after the fact.

## Environmental Responsibility

Accelerated innovation brings with it a growing environmental footprint, as training large AI models often consumes vast amounts of energy. AI-driven IoT systems can optimize energy consumption in smart cities, but ironically, the data centers powering those same systems may increase overall carbon emissions if not carefully managed. Balancing the benefits of AI against its ecological costs highlights yet another sphere where governance, corporate responsibility, and public awareness converge.

## The Opportunity for Human Flourishing

In spite of these perils, AI remains uniquely positioned to help us address urgent global challenges. From climate modeling to pandemic forecasting, from democratizing education to revolutionizing healthcare, AI can act as a force multiplier for human ingenuity. As we shape it responsibly, the technology can spark new forms of cultural expression, collaborative research, and global solidarity.

## Expanding Knowledge and Creativity

AI-driven tools already empower writers to explore bolder narratives, musicians to compose immersive soundscapes, and scientists to spot patterns in cosmic data sets. With each new generation of algorithms, creative and intellectual frontiers expand. This amplifies human potential artists gain fresh palettes, researchers get sharper instruments, and activists wield better analytical tools to highlight inequities or propose data-driven solutions.

## Addressing Global Inequalities

Properly channeled, AI could level the playing field for communities often left behind by globalization. Low-cost AI diagnostics might offer remote populations first-rate medical insights. Automated translations could preserve linguistic diversity while bridging cultural gaps. Autonomous drones and logistical optimizers could ensure essential supplies reach disaster-stricken regions rapidly. The vision of inclusive AI is not naive; it's grounded in the understanding that deliberate policy, equitable data collection, and robust community participation can direct AI's benefits where they're needed most.

## Revitalizing Civic Life

As we integrate AI, we also have the chance to reimagine social institutions. Education systems might shift away from rote memorization toward problem-solving and empathy training, because so many informational tasks can be automated. Workplaces might evolve into collaborative spaces where routine tasks are delegated to algorithms, freeing humans for strategic thinking or interpersonal connections. These transformations could reignite civic engagement if people feel they have a say in how AI is implemented fostering a sense of collective ownership rather than passive acceptance of technological dictates.

## Collective Responsibility and Privilege

We, as a collective, have both the responsibility and the privilege to direct these technologies toward inclusive, ethical, and inspiring outcomes. To see AI solely as a threat ignores its capacity to amplify positive human endeavors. Conversely, viewing it merely as a beneficial tool overlooks the historical patterns of oppression and exploitation that can be magnified by powerful new technologies.

## Individual Roles

Students and Lifelong Learners: By acquiring AI literacy, you can question hype, spot biases, and contribute ideas that shape the technology more inclusively.

Professionals Across Industries: Lawyers ensure fair legal frameworks, healthcare workers integrate AI ethically in patient care, journalists hold tech companies accountable, and so on.

Community Activists: Local data projects, civic forums, and advocacy for responsible AI policies amplify grassroots perspectives that might otherwise be overshadowed by corporate lobbies or government bureaucracies.

## Institutional Leadership

Tech Companies: They can design inclusive products, open-source critical research, and proactively address biases. This not only earns user trust but also demonstrates long-term foresight.

Governments: Crafting balanced legislation that fosters innovation while safeguarding civil liberties is crucial. Governments must also support public education on AI, ensuring more equitable opportunities in an increasingly automated world.

Academia and Nonprofits: By cultivating interdisciplinary research teams mixing computer science, sociology, philosophy, and beyond these institutions can illuminate AI's broader impacts. They can propose ethical frameworks, highlight overlooked communities, and drive dialogues that merge theory with real-world accountability.

## International Cooperation

Global Standards: We have already seen how Europe's data protection laws echo worldwide. Similarly, risk-based AI frameworks could inspire or pressure other regions to adopt parallel measures.

Shared Research Initiatives: Collaborative projects that address climate change, pandemic readiness, or humanitarian relief benefit from open datasets, open-source code, and shared methodologies across borders. This fosters a sense of shared destiny rather than a fragmented arms race.

**Looking Ahead: The Endless Frontier**

The arrival of Neuralink's advanced brain-computer interfaces and OpenAI's o3 model reminds us that the frontier of innovation is always evolving. Today's prototypes become tomorrow's everyday realities, and the line between human cognition and machine intelligence could blur more than we ever thought possible. Implantable devices might eventually bridge human senses with AI insights, or self-improving models might crack scientific puzzles that have long eluded us. In parallel, we'll see new ethical conundrums surface about autonomy, mental privacy, and the essence of consciousness.

# From Potential to Practice

The real challenge lies in turning ambitious potential into practical progress that people can feel in their local communities safer streets, improved healthcare, better educational resources, environmental rejuvenation, and broader cultural expression. This involves not just breakthroughs in labs but also robust support from policymakers, inclusive financing models, and public willingness to learn and adapt.

## Navigating Complexity

The path forward requires humility. AI interacts with countless variables economic markets, psychological biases, cultural values, international politics, and ecological constraints. No single blueprint will suffice. Instead, ongoing dialogue, experimentation, and revision will guide us. This iterative approach can harmonize local autonomy with global collaboration, ensuring that solutions remain sensitive to cultural and ethical differences.

## The Privilege of Choice

Ultimately, forging AI's future is both a responsibility and a privilege. Many past eras lacked the infrastructure or communicative reach to shape transformative technologies collectively. Today, we have social media platforms, educational tools, and open-source networks that allow even a single voice to resonate widely. Harnessing these capabilities responsibly means acknowledging that we can indeed steer the course. We don't have to accept AI applications that violate privacy or exacerbate social hierarchies as inevitable. We can demand better, strive for fairness and inclusivity, and build alliances that reflect those convictions.

## A Final Call to Action

By reading this book, you've already taken a step toward deepening your understanding of AI's complexities and promise.

The question now is how you will translate that knowledge into meaningful engagement. Will you champion ethical standards at your workplace, participate in local AI governance discussions, or join open-source initiatives tackling community challenges? Will you support policies that ensure the equitable distribution of AI's benefits, or perhaps create art that reflects the multifaceted relationship between humans and intelligent machines?

Whatever your path, remember that AI evolves through a network of human choices made by researchers, entrepreneurs, politicians, teachers, journalists, and ordinary citizens. Your viewpoint, your activism, and your creative insights matter. By acting collectively, we can ensure that advanced brain-computer interfaces or next-generation AI models enrich, rather than diminish, our shared humanity. Each of us holds a piece of the puzzle, and by assembling them together, we can shape a world where responsible innovation thrives, nurtured by an unwavering commitment to human dignity, equity, and the flourishing of all life on this planet.

May our collective journey continue with curiosity, compassion, and an unshakable resolve to make AI a genuine partner in creating a more just, inspired, and sustainable future.

## Afterword

As this manuscript concludes, fresh headlines about AI swirl in the media announcements of more advanced large language models, robotics innovations that push the boundaries of automation, and fierce debates in parliaments and tech conferences about regulating AI. While the specifics may shift, the guiding principles endure: centering human well-being, ensuring justice, and harnessing technological power for sustainable prosperity. My hope is that you've found in these chapters not just knowledge, but also inspiration and a sense of responsibility so that wherever AI goes next, we shape it with wisdom, empathy, and collaboration.

# Appendices

## Glossary of Key Terms

AI (Artificial Intelligence): Technologies and methods that enable machines to replicate tasks requiring human-like reasoning, perception, and learning.

Machine Learning (ML): A subset of AI focused on algorithms that learn patterns from data rather than following strictly programmed instructions.

Deep Learning: An advanced branch of ML involving large, multi-layered neural networks capable of abstracting high-level features from raw data.

Neural Networks: Algorithms inspired by the structure of biological brains, comprising interconnected artificial neurons that process numerical information.

AGI (Artificial General Intelligence): A hypothetical level of AI capable of performing any intellectual task that a human can, in a generalizable manner.

Reinforcement Learning: A learning paradigm where an AI agent optimizes actions through rewards or penalties in an environment, often achieving superhuman performance in games and simulations.

Transformers: A neural network architecture (e.g., GPT-4) that uses attention mechanisms to process sequential input (like text) efficiently and effectively.

IIT (Integrated Information Theory): A theoretical framework suggesting consciousness corresponds to the integration of information within a system.

Black Box Problem: A critique of complex models whose internal logic is not easily interpretable by humans, making accountability and transparency challenging.

Algorithmic Bias: Systematic errors in AI outputs that stem from biased data or flawed model assumptions, potentially leading to discrimination in real-world applications.

Brain-Computer Interface (BCI): A technology that enables direct communication between the human brain and external devices, often employing implantable or wearable sensors.

o3 Model: An advanced AI architecture developed by OpenAI, designed to mirror human cognitive processes in reasoning, problem-solving, and creative tasks.

Javad Yahaghi

# Selected Bibliography and Resource List

- Russell, Stuart & Norvig, Peter. *Artificial Intelligence: A Modern Approach (4th Edition)*. Pearson, 2020.
- Goodfellow, Ian; Bengio, Yoshua; Courville, Aaron. *Deep Learning*. MIT Press, 2016.
- Bostrom, Nick. *Superintelligence: Paths, Dangers, Strategies*. Oxford University Press, 2014.
- DeepMind's AlphaFold: Official website & research publications on protein folding breakthroughs.
- IEEE: *Ethically Aligned Design: Global Initiatives on the Ethics of Autonomous and Intelligent Systems.*
- UNESCO: *Recommendation on the Ethics of Artificial Intelligence*, 2021.
- EU AI Act: Ongoing drafts and revisions from the European Commission.
- Stanford University's AI Index: Annual report tracking global AI research, investment, and societal impact.
- MIT Technology Review: Regularly publishes AI breakthroughs, policy discussions, and technical analyses.
- OpenAI, Google AI Blog, Microsoft Research: Leading platforms documenting recent AI techniques, open-source projects, and scientific papers.
- Neuralink Official Site and Research Papers: Updates on clinical trials, implant technologies, and breakthroughs in brain-computer interfaces.
- OpenAI o3 Documentation: Technical whitepapers and community forums detailing the o3 model's architecture, use cases, and ongoing development.

# Index

- Adaptive Learning: 36–44
- AGI (Artificial General Intelligence): 24–31, 106
- Algorithmic Bias: 80–82
- BCI (Brain-Computer Interface): 145–148
- Deep Learning: 7–8, 13-14, 27–38
- Ethical Frameworks: 77–85, 175–176
- EU AI Act: 86–87, 113–114, 177
- Financial Applications: 45–46
- Healthcare Transformations: 33–38
- Machine Consciousness: 69–676
- Machine Learning: 13–14
- Neuralink: 145–149
- Quantum Computing: 138–139
- Singularity: 98–105
- Smart Cities: 124–134
- Sustainability: 124–134
- OpenAI o3 Model: 184–189
- Transportation: 57–58

www.ingramcontent.com/pod-product-compliance
Lightning Source LLC
Chambersburg PA
CBHW070352200326
41518CB00012B/2218